Particle Accelerators, Colliders, and the Story of High Energy Physics

Raghavan Jayakumar

Particle Accelerators, Colliders, and the Story of High Energy Physics

Charming the Cosmic Snake

 Springer

Raghavan Jayakumar
San Diego
CA, USA
raghavan.jayakumar@gmail.com

Credits: LHC image - Source: CERN
Cover images: Cosmic Snake eating its tail and surrounding the Large Hadron Collider

ISBN 978-3-642-22063-0   e-ISBN 978-3-642-22064-7
DOI 10.1007/978-3-642-22064-7
Springer Heidelberg Dordrecht London New York

Library of Congress Control Number: 2011939790

Printed on acid-free paper

Springer is part of Springer Science+Business Media (www.springer.com)

Preface

Soon I knew the craft of experimental physics...the sublime quality of patience – patience in accumulating data, patience with recalcitrant equipment...

Abdus Salam

I have had the privilege of being a wanderer in the diverse world of physics. That privilege vested in me an appreciation for all of physics, filling me with wonder. But when I became an accelerator physicist in 1990, I experienced the true joy and fulfillment that only something truly sublime could bring. I saw how physics principles, ranging in complexity from simple, elegant laws, to hard to understand complex notions, had to come together in defining the mission of large physics facilities. I saw the true nature of spectacular experiments, facilities, and cutting-edge technologies that must be developed and built, to go on a quest for the ultimate discoveries in physics. The enormity of the scientific, engineering, and management endeavor fascinated me, while the politics vexed me. I felt the excitement of a child who gets on a ride in an amusement park for the first time. The Large Hadron Collider came on line, sparking the imagination of physicists and the general public alike. This increased my pulse rate. I felt I had to tell someone about what accelerators–colliders–detectors are, and how they came about. Hence the book.

I also feel that the public should weigh in strongly on the decisions to pursue the increasingly complex and expensive physics experiments. This is only possible if they understand the concepts and implications, to the depth that is required in order that no one can manipulate their opinions. So while this book has a fair amount of mathematical descriptions and what might be, to lay people, difficult concepts, it is still possible for a lay reader to skip such sections and appreciate the motivations, principles, and efforts required to build particle accelerators and detectors. In this context, readers are urged to take note of the large number of occasions when a scientific invention intended only for a specific experiment resulted in a major technological application. Examples range from the Cathode Ray Tube to X-rays to nuclear power to the Internet.

Another motivation for writing this book is to provide reading material for undergraduate science and engineering students. A good familiarity with this branch of physics that connects to the greatest aspirations of physicists would increase their love for science and their commitment to pursue it, either in their career or as a hobby. Therefore, this book has been written at a somewhat higher level than popular physics and includes several detailed derivations. In all cases, there has been an attempt to provide as complete a description as possible on a given topic, while limiting it to undergraduate level math. Overall, the book is written to provide a comprehensive understanding of the chain of developments. Details and some math are added where necessary, to clarify the concepts. With this approach, it is hoped that undergraduate students would find it a satisfying read while enriching their knowledge in the field of physics. I hope this book will also provide a context to their course work, and firm up their learning.

The evolution of accelerators has been rapid, and makes for a fascinating story. Over the last several decades, the development of physics itself has been closely linked to this evolution. Therefore, the chapters describe the developments in accelerators and detectors through linkages to fundamental physics goals of the time. The fascinating history of very humble beginnings with terrestrial radioactive sources of particles to the modern day powerful colliders, is told through narratives that include human interest stories as well as physics content. In addition to photographs and illustrations, cartoon illustrations are included to provide analogies and to lighten up these heavy descriptions. All along, attempt has been made to convey the excitement of the discoveries, and the admiration and wonder that the author feels when thinking about the science and the scientists. When experimental facilities grew very large, and governments became the only possible sources of funding, the history of high-energy particle experiments, like most branches of big science, became permeated with politics, for good or bad. The narration includes cases where politics was instrumental in shaping the outcome.

The first three chapters provide the background and motivation that led to the discovery of the first particle accelerators. The first two chapters trace the development of the notion of atoms, the discovery of electromagnetism, and the first discovery of a fundamental particle – the electron. These chapters also show the methodology of research that set the trend for future generations of researchers. Chapter 3 shows that man-made instruments such as the Cathode Ray Tube themselves spurred the investigation of new kinds of radiation – the alpha, beta, and gamma rays. The discovery of these rays led curious physicists to inquire into the structure of the atom itself, gaining a first glimpse into Nuclear Physics. The abiding discovery of cosmic rays and their measurement then rounded up the natural sources of particles that provided almost all of the physics clues of the early 1900s.

Chapter 4 then narrates the physics and history of the invention of the first accelerators, along with associated discoveries. The use of magnetic fields in building compact accelerators also led to new accelerator physics principles, sparking research into "betatron oscillations". Chapters 5–7 continue the narration, describing circular accelerators and the major discoveries that were made using these. The chapters describe how these machines grew from a size that would fit in the palm of a hand while providing tens of kilo electron volts of energy to "Rings of

Earth", which would enclose vast tracts of land and produce particle energies of trillions of electron volts. The physics of particle acceleration using radiofrequency fields and the behavior of particles in such a matrix of accelerating geometry is presented concisely. Simultaneously, the story of the centers of particle accelerators and high-energy physics research, such as the Brookhaven Laboratory and the CERN, is told. Chapter 8 describes the superconducting circular accelerators and the cutting-edge technologies that make them possible. Chapter 9 continues the story of the fascinating evolution of linear accelerators, which was interrupted by the development of circular accelerators, but set back on course by new physics and technology. Chapter 10 is relatively more difficult, dealing with the physics that is on the minds of researchers of today. Here is a glimpse into the rich and complex physics that encompasses present understanding, and the questions that remain unanswered. It departs from the narration of the previous chapters in that little mention is made of particle accelerators. Instead, the physics discoveries, many of which were theoretical, are presented as a foreground to present quests in the frontiers of physics. Following this physics, the engineering, technology, and management backgrounds of particle colliders and detectors are given in Chaps. 11 and 12. Finally, Chap. 13 describes the 7 TeV on 7 TeV Large Hadron Collider, the pride of the physics nation. Included in this chapter are the promise of this fantastic experimental facility and explanations on some of the myths that surround it. In this way, the book is hopefully a journey through the history of particle physics and accelerators, leading to the present. Through this journey, it is hoped that readers would acquire adequate familiarity so that they would recognize the subcomponents of accelerators and detectors if they were to visit a facility.

The writing of this book has itself been a journey and a personal quest for me. Hence I would like to acknowledge with gratitude those who helped me along the way. Dr. Ramon Khanna has been a critical editor who, along with the editorial team, guided the content and reviewed the material critically. I wish to thank my wife Suhasini Jayakumar for her critique and help with the narration and editing. I also acknowledge, with immense gratitude, all the people and institutions that gave permission to use their illustrations and photographs and gave helpful comments.

R. Jayakumar

Special Acknowledgement

A few years back, David Orrell and I started writing a similar book together, but we could not continue because of circumstances. Chapter 3 is drawn from this collaboration and David Orrell is a significant contributor to Chapter 3. His superior writing style is evident in this Chapter and I gratefully acknowledge his contribution.

R. Jayakumar

Contents

Chapter 1
The Expanding Universe of Particles

Student of Democritus: My teachers Leucippus and Democritus firmly believed that matter consists of atoms, the smallest, indivisible building blocks.
Aristotle: Those guys were full of it. Actually, all matter is a combination of four elements – Fire, Water, Air and Earth. There can be nothing like atoms.
Student: And why not?
Aristotle: Because matter has to be continuous. If it were to consist of atoms then there would be gaps and the gaps would have vacuum.
Student: Why can't there be vacuum?
Aristotle: Silly, that would mean that an object can then travel at infinite speed without any resistance and that is not permitted.
Student: Hmmm...
(An imagined dialog in Greece between Aristotle, a Greek philosopher around 350 BC, and a student of Democritus.)

Aristotle called the understanding of "*phusis*" (nature) the First Philosophy. From the earliest recorded inquiries on the nature of nature, by philosophers in the city of Miletus in Greece in sixth century BC to today's research by theorists and experimentalists around the world, the First Philosophy's primary question has been – what does matter consist of. Aristotle was such an influential thinker that any notion on the building blocks of nature other than his own, had no chance of making headway. Contemporaneously with Greeks, Indian philosophers had conceived the *panchabhoota*s, the five elements that included ether as the fifth element. But there is an indication in the Hindu scripture Bhagawat Gita that the atomist (*Anu*) concept of matter was already in coinage in India, in as early as fifth century BC. While the Indian concepts remained in the religious domain, Democritus described the role of atoms in composing matter (Fig. 1.1). He also explained that atoms wander off from an object and give us aromas and smells. So, such philosophers belonging to this atomist philosophy laid the foundation for admitting a physical world that was outside human conventions, sensations, gods, and faith. Yet, all these were mere theories and there were few "physics" experiments to prove or disprove any of the prevailing concepts. The experiments were limited to observing terrestrial and

R. Jayakumar, *Particle Accelerators, Colliders, and the Story of High Energy Physics*, DOI 10.1007/978-3-642-22064-7_1, © Springer-Verlag Berlin Heidelberg 2012

Fig. 1.1 Over millennia, identity of fundamental particles has been changing, man-made accelerators might be bringing a closure to this issue

celestial phenomena. Looking back, it is astounding that the intellectual giants of those times were unable to devise observational experiments to settle issues. Two facts demonstrate the resulting hold of Aristotle's ideas for two millennia – (1) it was wrongly believed that plant mass was gathered from the soil through the roots, until in 1700s Wiegmann and Polstroff demonstrated that most of it came from the air. (2) Aristotle had believed that heavier objects fall faster than lighter one, until Galileo proved it otherwise in an elegant inclined plane experiment in the year 1604, perhaps one of the first real physics experiments.

The Elements Abound and Atomic View Prevails

The "four-element" basis of matter started losing out in the face of new discoveries from the late seventeenth century onward. A Parisian chemist Antoine Lavoisier deserved much of the credit for establishing the atomic basis of matter, though he himself had kept his mind open on it. The Greek theory was that flammable materials contain the so-called *phlogiston* that was released when the materials burned, leaving behind the dephlogisticated material *calx*. Because nothing burned without air, it was assumed that air absorbed all the phlogiston. Performing an experiment in 1777, Lavoisier showed that not only did burning mercury (heated in the presence of air) gain weight even though it lost phlogiston, but increase in weight of the converted mercury (HgO) was also accompanied by a corresponding reduction in the amount of air, which should have gained weight if it were absorbing phlogiston. The law of conservation of mass too was established. But when the substance was reheated strongly, the dephlogisticated air was released back (HgO was reduced), which is impossible from the phlogiston theory. In 1785, he also demonstrated that oxygen (which means becoming sharp), so named by himself, combined with the inflammable air hydrogen to produce water. Unfortunately, Lavoisier did not live to enjoy his new findings and extend them. French revolutionaries guillotined him in 1794. Lagrange is quoted as saying *"A moment was all that was necessary in which to strike off this head, and probably a hundred years will not be sufficient to produce another like it."* Some would, however, say that he deserved it because he was not a man of honor and he took other people's scientific work without acknowledgment (*Famous Chemists* by Sir William Tildon, George Routledge & Sons, E.P. Dutton & Co., New York, London, 1921).

John Dalton, a meteorologist and a physicist, continued where Lavoisier left off. A confirmed atomist, in the strict sense of the term, not only did he hold that all matter consisted only of atoms, but he also believed that atoms were indeed indivisible. His first remarkable results came in 1803, when he carefully measured the ratio of atomic weights of 6 elements to that of hydrogen by observing the weight of the element consumed in a reaction. Here, he assumed that one atom combined with only one atom, and such assumptions gave incorrect answers and he faced difficulties. (As we know today, compounds are formed by the combination of number of atoms of each element according to their chemical valence). Amedeo Avagadro proved in 1811 that equal volumes of gas at a given temperature and pressure contained equal number of particles, which was assumed to be atoms (later we would know that it is molecules). This resolved many difficulties with Dalton's theory. Once the concept of atoms took hold, the concept that matter is constituted from individual elements that consisted of atoms was admitted and indeed, new elements were discovered. By 1863, the universe was suddenly alive with 56 elements, an element being discovered every year. In a major display of pattern recognition, in 1869 Dimitri Mendeleyev constructed the periodic table based on chemical valence and atomic weights of then-known elements. While Lothar Meyer also constructed an identical table independently, Mendeleyev gets the credit, because he went on to predict the existence of yet undiscovered elements similar to boron, aluminum, and silicon. The amazing fact shown by Mendeleyev's periodic table that elements ordered themselves according to Dalton's atomic weights, became the proof of an atomic nature of matter. Democritus smiled in his grave.

Chapter 2
The Spark that Broke the Atom

We are once again brought back to Miletus, Greece, in 585 BC, where the respected scientist – philosopher Thales, discovered electricity by rubbing fur against amber and he could even produce a spark. Since this discovery, electricity remained, more or less, a matter of curiosity and an unresearched phenomenon, until the eighteenth century, when discovery of magnetic properties of lodestone (magnetic materials found in nature) by William Gilbert and his subsequent detailed comparison of electricity and magnetism were made. Preliminary experiments such as those by C.F. du Fay on two different forms of electricity (now known to be positive and negative charges) and by Benjamin Franklin on lightning kept electricity as an interesting research field, but perhaps only still a curiosity. However, once Alessandro Volta made a reliable source of electricity by constructing a battery using alternate layers of zinc and copper immersed in an electrolyte, and Michael Faraday invented the electric motor that could replace animal and manual labor in 1821, the field of electricity had made it into the engineering field and the research on electricity and magnetism became a very profitable necessity. Following the demonstration of the connection between electricity and magnetism by Hans Oersted and Andre-Marie Ampere, and analytical descriptions of circuit currents and voltages by George Ohm, the field became a bread-and-butter activity of scientists around the world. From here on and rightly so, scientific research was concerned with the field of electromagnetism for a long time. Even today, majority of inventions and tools resulted from such research.

In the 1830s, Faraday and Ampere showed that time-varying magnetic field induced electricity and time-varying electric field induced magnetic field. This then implied the possibility of a new kind of wave to enable the coexistence of alternating electric and magnetic fields. In 1864, James Clarke Maxwell settled the issue with his discovery of the equations governing electromagnetism. These equations, which are the equivalent of Newton's laws of motion but for electromagnetism, describe the function of most of the electromagnetic devices we use today. As we shall see, even primary descriptions in many topics in this book require an appreciation of Maxwell's equations.

R. Jayakumar, *Particle Accelerators, Colliders, and the Story of High Energy Physics*, DOI 10.1007/978-3-642-22064-7_2, © Springer-Verlag Berlin Heidelberg 2012

Faraday's law is encoded in Maxwell's equation as

$$\nabla \times E = -dB/dt, \tag{2.1}$$

where t is time, E is the electric field, and B is the magnetic field induction. The left-hand side gives the "curl" operation of electric field. This states that the variation (and therefore the presence) of electric field in space (curl operator – describing variation with respect to spatial directions, perpendicular to the vector quantity) is associated with time variation of a magnetic field (which is perpendicular to the electric field). This is also the law of induction (see, for example, betatrons) which is the basis of transformers, where a changing magnetic field due to changing current in the primary induces voltage in the secondary of the transformer. Ampere's discovery of magnetic field generation by an electric current and/or time-varying electric field is described by the equation

$$\nabla \times B = \mu_0 (J + \varepsilon_0 dE/dt). \tag{2.2}$$

(μ_0 and ε_0 are the so-called vacuum permeability and permittivity physical constants, respectively.) This complements the previous equation and states that the spatial variation (existence) of magnetic field is associated with time varying of electric field (which is perpendicular to the magnetic field) and any electric currents driven by electric fields. Obviously, the latter forms the principle behind magnets, while the former is most associated with capacitors. Two other equations,

$$\nabla \cdot E = \rho/\varepsilon_0; \nabla \cdot B = 0, \tag{2.3}$$

describe the electric field variation around a charge density and the fact that magnetic field lines are always closed, respectively (The closing of magnetic field lines implies that magnetic charges do not exist. However, existence or nonexistence of magnetic monopoles, equivalent to an isolated positive or negative electric charge, remains uncertain.). Strikingly, combining these equations one gets the wave equation

$$\nabla^2 E = \frac{1}{c^2} \frac{\partial^2 E}{\partial t^2}, \tag{2.4}$$

where $c = \sqrt{1/\mu_0 \varepsilon_0}$ is the speed of the wave. An identical expression for the wave magnetic field is also obtained. In an astounding discovery, when the speed of light was measured, it turned out to be the same as the speed of the electromagnetic wave. Since then, we have known that all light are electromagnetic waves capable of traveling in vacuum. We also know now that electromagnetic waves span all values of frequency feasible within nature – radio frequency waves to gamma rays.

The Electricity Carrier

All the scientists were convinced now that something precipitated to charge something electrically and ran around to carry a current. The varying ability of this carrier to run through material made the difference between good conductors and bad conductors. They even knew that chemical reaction (valence of elements) and electrolytic properties were connected to the nature of charge on atoms of elements, but did not know what this charge carrier might be. Like many physics results, the discoveries had to await technological developments, which, in turn, depended on previous scientific discoveries. While this symbiosis between science and technology is critical even today, it is only vaguely understood by general public and barely acknowledged by technologists.

It has long been known that when glass tubes are evacuated, one could pass electric current through the evacuated space in the form of arcs from a metal cathode to a metal anode. Then in the 1850s, Heinrich Geissler improved the vacuum techniques and showed that one could get glowing discharges between electrodes in an evacuated tube. In 1879, William Crookes obtained a vacuum of better than 10^{-4} mmHg and found that the tube became dark, filled with the so-called Faraday Dark Space. When other experimenters used this tube, they found that at one point, the glass behind the anode was fluorescing and painting the glass with phosphorescent coating made a very clear illuminated spot. It was surmised that a "cathode ray" was being emitted from the negative electrode and had been accelerated by the anode through the dark space. Crookes called it "radiant matter." Some others thought that these charge carriers were a form of electromagnetic radiation (ethereal disturbance), because Hertz found that the rays penetrated gold foil. It was then inconceivable that solid particles could pass through solid metal.

Fig. 2.1 J.J. Thomson's Cathode Ray Tube with deflecting electrodes and Helmholtz coils. Image No. 10324719 © Science Museum/Science & Society Picture Library

Others such as Shuster thought this was a charged atom (after all, in those days, the atom was indeed the atom, the indivisible).

The First Fundamental Particle in the First Philosophy: The Electron

Crookes did believe that the rays were negatively charged, because they were attracted by the anode. But Hertz found that these rays were not deflected by electrically charged plates, as they should have been if these were charged particles. The fascinating experiments with the marvelous discharges went on with ever improving techniques. In 1897, with these advances, Joseph John Thomson at the Cavendish laboratory in Cambridge not only finally proved that the rays were made of particles, but he also managed to determine their mass.

When J.J. Thomson repeated Hertz's experiment at a very low pressure, the rays were indeed deflected by charged plates. It turned out that in Hertz's experiment, the tube pressure was not low enough and the remaining ionized gas shielded out the electric field from the plates. At this time, the physics of electromagnetic phenomena and charge motion were well established and therefore, Thomson could make measurements and calculate the properties of the current (charge) carrier from those measurements. In his defining experiment, Thomson measured the ratio of the particle charge to its mass.

Thomson determined the charge-to-mass ratio by applying an electric field (perpendicular to the direction of cathode ray particle moving with velocity v) to bend the path of the particles, and compensated the deflection by applying a magnetic field and brought the ray back to the direction it was traveling before encountering the electric field (see Figs. 2.1 and 2.2). This used the fact that force in a magnetic field depends on the charge and the velocity with which the particle travels (The magnetic field direction has to be perpendicular to the direction of the ray as well as the deflection). Then, for a given force, the acceleration (and the velocity gained in the transverse direction after traveling certain distance in the field) depends on the mass.

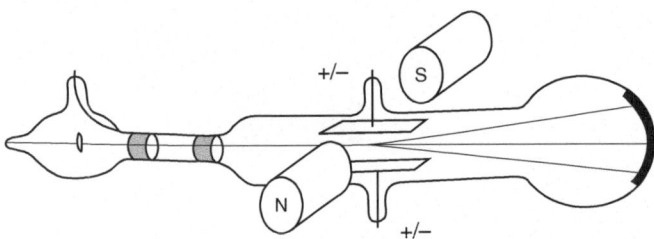

Fig. 2.2 A schematic of arrangement used by J.J. Thomson to measure electron charge-to-mass ratio

A particle with a charge $-e$ experiences a force $-eE$ in an electric field E. If a magnetic field B is applied in a direction such as to compensate the bending, then the force is eE (direction compensating). When the deflections cancel,

$$eE = -evB, \quad \text{so that} \quad v = -E/B. \tag{2.5}$$

Now, the radius of deflection of a charge was even then known to be equal to $mv/(qB)$, where q, m, and v are the charge, mass, and velocity of the ray particle and B is the magnetic field. Therefore, the deflection when only magnetic field is applied is proportional to the ratio of mass to charge. This results in the relationship which gives the ratio of particle charge to mass as

$$e/m = (\theta E/LB^2) \tag{2.6}$$

Thomson also measured the deflection angle θ of the ray when only magnetic field was present (by the arrangement described above, this is also the deflection caused by the electric field only). Each of the quantities on the right-hand side of (2.6) (where L is the length of the magnetic field region) was measured by Thomson. When he calculated it, he found the ratio of charge to mass to be very high, 2,000 times that of the hydrogen ion (positively charged hydrogen). This meant that the cathode ray particle was very light, 2,000 times lighter than that of the hydrogen ion, or it had a charge 2,000 times larger than hydrogen. From (2.5), he also determined the speed of the cathode ray particle to be about 100,000 km/s, a third of the speed of light! From the estimates of energy absorbed, he suspected that the cathode ray particle was much lighter than hydrogen. Philipp Lenard conclusively showed that cathode rays were lighter rather than being highly charged, by studying their passage through various gases (later, Robert Milliken would measure the charge of the electron to high accuracy to confirm this). Thomson concluded that the particle, whatever it was, appeared to "form a part of all kinds of matter under the most diverse conditions; it seems natural therefore to regard it as one of the bricks of which atoms are built up." He called these particles "electrons," as befitted the carrier of electricity. Thomson also invented the new method of "detecting" and measuring particles using electric and magnetic fields, which is used even today in mass spectrometers for selecting particles with specific velocities. The Cathode Ray Tube pioneered by Crookes and others forms the basis for television tubes.

When Thomson announced the discovery of electron on April 30, 1897 to an audience, they thought he was pulling their legs. This startling conclusion laid the atomic basis of matter as the indivisible building block to rest. The electron would become the first and enduring fundamental particle, and Thomson became the celebrated pioneering discoverer of a fundamental particle. To this day, there is no evidence that the electron has any underlying structure. Electron behavior influences every moment of this universe and our own lives.

Chapter 3
Nature's Own Accelerator

Thomson's first idea for how the atom combined positive and negative electrical charge – known as the plum pudding model – was that it consisted of a positively charged lump of mass studded through with the negatively charged electrons. Experiments using a kind of particle accelerator would later prove this model wrong – but here the accelerator was not man made.

While people tend to think of radiation as being man made, our bodies are constantly bombarded with naturally occurring radiation, both from terrestrial radioactivity and solar and cosmic rays. The average naturally occurring radiation is about 240 milliRem per year, about 500 times larger than the average dose released in nuclear testing and in nuclear power station accidents and about five times larger than dosages received in medical and dental treatments. (One would have to receive about 100 Rem per year to have a 5% chance of developing cancer later in life.) The ground contains radioactive substances like uranium, and the heavens contain the radiation of exploding stars, radiation from the Sun, and even echoes of the Big Bang. The detection of such radiation by Victorian scientists was linked to an imaging method that was just becoming popular with the general public: photography.

X-ray Eyes

In 1895, the 50-year-old German physicist Wilhelm Röntgen was experimenting with his cathode ray tube, which he had covered with a black card in order to block out its glow. On November 8, he left some tightly wrapped, unexposed photographic plates near the tube. When he later pulled out the plates for use, to his puzzlement, he found that they had been fogged, as if exposed to light. He also noticed that a sheet of paper coated with barium platinocyanide – which fluoresces when exposed to light – glowed in the dark when brought near the machine. It would seem that some kind of invisible light was being emitted from the machine, and it occurred to Röntgen that if this was the case, it should be possible to use it to

R. Jayakumar, *Particle Accelerators, Colliders, and the Story of High Energy Physics*,   11
DOI 10.1007/978-3-642-22064-7_3, © Springer-Verlag Berlin Heidelberg 2012

take a picture. He first tried placing solid objects in front of a photographic plate and exposing them to the radiation. He found that objects were transparent to the radiation, to a degree that depended on their thickness. He then asked his wife to hold her hand steady over a photographic plate. The result was the world's first X-ray picture. The X-rays, as Röntgen named as whatever they were, passed easily through his wife's flesh, less easily through bone, and not at all through her wedding ring (Fig. 3.1).

Though these mysterious rays were first noticed by another German researcher Johann Hittorf in 1875 and even quantitatively measured by Ukraine born Ivan Pulyui in 1886, Röntgen is credited with the discovery and the preliminary under-standing of it. The discovery of X-rays immediately caused a sensation, and a fair amount of titillation, among the general public. Victorian women began to worry about mad scientists taking pictures of them under their clothing. But the scientists were more interested in determining the nature of the rays. It seemed that they were not the same as cathode rays – for one thing, they were unaffected by electric or magnetic fields – but scattered out in all directions when the cathode rays hit a material object, such as the end of the tube.

We now know that the X-rays are just a form of electromagnetic radiation and are emitted as a result of two phenomena – (1) X-ray fluorescence, in which electrons impinging on a target atom excite the orbital electrons of the atom, which then decay to the ground state releasing the quantum of energy absorbed from the electrons, as an X-ray photon. This results in a discrete spectrum of X-rays.

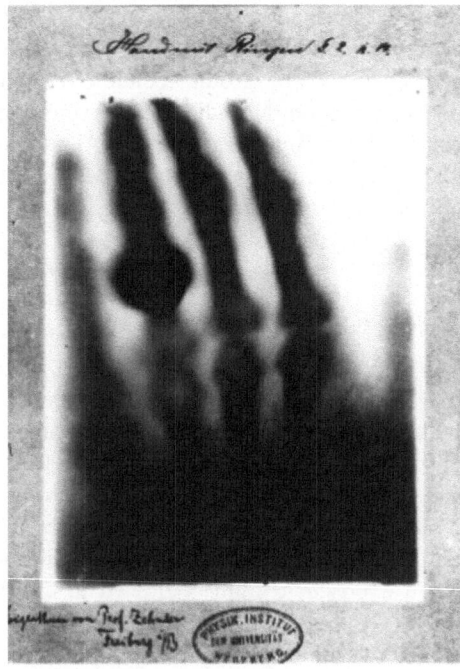

Fig. 3.1 The first X-ray picture of a human body part. Picture of his wife Anna Bertha's hand with a wedding ring, taken by Rontgen in 1895, Physik Institute, University of Freiberg, Germany

(2) X-rays are also emitted due to Bremsstrahlung radiation, which is emitted when the impinging electron is slowed down in the vicinity of atoms (particularly ones with high atomic number) and this results in a continuous spectrum of X-rays and other radiation. These are the type of emissions, common in many light sources, such as arc lamps and Sun. In general, X-rays are emitted when high-energy electrons interact with matter.

Invisible Rays Out of Nowhere: Radioactivity

In January 1896, the French physicist Henri Becquerel, who worked at the French Museum of Natural History in Paris, heard of Röntgen's discovery at a meeting of the French Academy of Sciences. He wondered whether phosphorescent materials – which glow in a similar manner to a cathode ray tube – also produced X-rays. He immediately tried a simple experiment. In earlier work with his father, he had discovered that uranium salts glow in the dark (emitting visible light) after being exposed to sunlight. He wrapped a photographic plate in black paper, placed a sample of uranium salts on top of it, and left it in the sun so the salts would absorb sun's rays and then glow in the dark. Sure enough, the salts exposed the plates, proving that they too produced rays that could pass through the black paper.

At the end of February, Becquerel tried repeating the experiment, this time with a small metal cross between the salts and the plate to see if it would leave an outline. However the sun didn't cooperate, and after several days of cloudy weather, he either grew bored or decided to perform a control experiment and developed the plates anyway. He was amazed to find that the salts had exposed the plate and left the outline of the metal cross, even though they weren't glowing (not emitting visible light). He concluded that the material itself was emitting an invisible ray, which he assumed were again X-rays. Yet, the puzzling aspect of the phenomenon was that it seemed to violate conservation of energy. Where was the energy coming from to produce these rays, if it wasn't originally absorbed from the sun? The answer to this question owed much to a young Polish couple living in France.

Marya Sklodowska was born in 1867 in Warsaw, which at the time belonged to the Russian part of a divided Poland. She became interested in chemistry and physics, but women were not admitted to universities in Warsaw. Instead she joined a group of patriotic Polish youths who would gather secretly to avoid the Russian Czar's police, and take turns giving lectures on a range of topics. In 1891, she moved to France to study at the Sorbonne. Life in Paris on her limited funds was not easy. As she later wrote, "The room I lived in was in a garret, very cold in winter, for it was insufficiently heated by a small stove which often lacked coal. During a particularly rigorous winter, it was not unusual for the water to freeze in the basin in the night; to be able to sleep I was obliged to pile all my clothes on the bedcovers." (Marie Curie, A life by Susan Quinn, Persius, Washington D.C. (1995), p. 91.) She was luckier in love, and in 1895 married Pierre Curie. And 2 years later, Marie Curie – as she was now known – became one of the first women in Europe to embark on a Ph.D. Perhaps out of an understandable desire to find an infinite source of heat, she chose as her research topic, the mysterious phenomenon of radioactivity (a name she coined), by which materials like uranium seemed to produce energy from nothing.

Curie soon discovered that the uranium ore, known as pitchblende, was actually more radioactive than pure uranium itself. She concluded that it must contain some other highly radioactive substance. Working together with her husband Pierre, she eventually succeeded in isolating not one but two new sources of radiation. One they called radium, and the other, in a political gesture to their homeland which was still under Russian rule, polonium. Both were present in only trace quantities – a fraction of a gram in tonnes of pitchblende – but they were hundreds of times more radioactive than uranium. Pierre calculated, for example, that a lump of radium could heat more than its weight of water from freezing to boiling in 1 h – not just once, but over and over!

Their discoveries brought the Curies fame, but also disaster. In April 1906, Pierre was killed by a horse-drawn wagon after he slipped and fell in the street. He had been suffering from dizzy spells that were likely caused by radiation poisoning. Marie died in 1934 from leukemia, which was probably also the result of working with radioactive substances without proper protection. Her laboratory notebooks are still radioactive and are kept in a lead-lined vault. (The dangers of polonium were demonstrated more recently when it was used in the 2006 poisoning in London, of the former Russian spy Alexander Litvinenko. A tiny quantity put in his food or drink was enough to slowly destroy his internal organs.)

The Alphabet of Particles

Another scientist intrigued by the properties of radiation, but who apparently managed to avoid its harmful side effects, was the brilliant New Zealand physicist Ernest Rutherford, a dominant figure in experimental physics for several decades. A student at Cambridge under J.J. Thomson, he moved in 1898 to McGill University in Montreal. Soon, he dedicated his research to the study of Becquerel rays. There he noticed that he could identify the rays as a positively charged ray or negatively charged ray by passing them through a magnetic field. The charged particles bent away from their path in presence of a magnetic field, positive charge one way and the negative charge, the other way. A third ray left the magnetic field region without being affected by it. He named them alpha, beta, and gamma rays. Alpha rays (particles) had positive charge and a short range such that they were stopped even by a piece of paper – they couldn't punch their way out of a paper bag – while betas and gammas had more go. The beta rays had negative charge while gammas, unaffected by the field, had no charge. (We now know that alphas are the same as helium atoms, stripped of their two electrons, betas are high-energy electrons similar to cathode rays, and gammas are electromagnetic radiation with energies higher than X-rays.) So, Becquerel's wrapped plates were exposed to all the three rays. However, his plates were exposed to mostly beta and gamma radiation, since the alphas were mostly stopped by the wrapping paper.

The loss of energy with distance (x) of particles with energy E (energy loss per unit distance traveled by the ray), a charge z and mass M through a medium with N atoms per unit volume, is given by the proportionality relation,

$$\frac{dE}{dx} \propto \frac{z^2 e^4 N}{E} \left(\frac{M}{m_e}\right) = \frac{S}{E}$$

where N is the particle density of the medium and e and m_e are the electron charge and mass. Denser is the medium (large the N), the particle slows down faster. The energy loss increases as the energy decreases and so the loss increases exponentially with distance and the range is exponentially reduced for large S. Since the alpha particle is heavy compared to electron (M/m_e is about 4,000), the energy loss rate is high for alphas. Alphas also have twice the charge of an electron and therefore, the energy loss is increased further by a factor of four, compared to the beta rays.

Working with the English chemist Frederick Soddy, Rutherford discovered how radioactivity could appear to produce an endless supply of heat for nothing. The reason was that it involved the change of one element into another. For example, when radium emits radioactive alpha particles, it transforms itself into radon gas (discovered by Friedrich Ernst Dorn). Every 1602 years, half the radium gets transformed in this way. After another similar period, half the remaining radium is lost, and so on. After a million years only a trace will remain. So the energy is being produced by the transformation of matter from one form to another, and eventually it runs out. Such transformation had long been the goal of alchemists, though the idea there was usually to change lead into gold. (Ironically, some high atomic number radionuclides actually decay into lead.) In 1908, Rutherford was awarded the Nobel Prize for chemistry! The award citation read: "..for his investigations into the disintegration of the elements, and the chemistry of radioactive substances." It is interesting to note that Rutherford's conclusion implied an understanding that energy and mass are equivalent, an idea that was just being proposed by Albert Einstein.

An example of alpha decay is of uranium into thorium and helium: $U_{92}^{238} = Th_{90}^{234} + He_2^4$. Today, we know that alpha decay is a quantum tunneling process (see Chap. 4) by which a metastable heavy nucleus with a larger (total) binding energy decays into a daughter product and an alpha particle. The daughter has a smaller binding energy than the parent. The binding energy can be thought of as a stored energy like the chemical energy in a fossil fuel and so the resulting difference in energy is mostly carried away by alpha particle. (Even though the parent nucleus has a larger total binding energy, the binding energy per nucleon – particles in the nucleus – is actually smaller than for the daughter nucleus, which is why the parent is more unstable.) Another interesting fact to note is that the conservation of the net zero momentum results in a narrow energy range of

5 MeV for the emitted alpha particles. The alpha particle slows down readily in matter because of its charge and mass and therefore survives only a few cm in air. Alpha decay is caused under the strong nuclear force and electromagnetic force (see Chap. 6). An example of the beta emitter is an isotope of carbon, which emits an electron and transmutes to nitrogen: $C_6^{14} = N_7^{14} + e^-$. While the alpha decay was relatively well understood, the physics of beta decay remained a mystery for considerable length of time. A new theory discovering a new force "the weak force" was needed to explain the process (see Chap. 10). The third type of emission from radioactive nuclei, named gamma rays by Rutherford, was identified by Paul Villard in 1900. These are electromagnetic radiation, often, from daughter products of alpha or beta decay, which leave a nucleus in an excited state and the nucleus returns to the ground state by emitting a photon, analogously to atomic radiation. While normally gamma rays have high energy, nuclear emissions may be at a wavelength as high as that of ultraviolet rays.

In 1907, Rutherford moved to the UK to take up the position as Chair of Physics at the University of Manchester, where he promptly set up a group to research the topic of radiation. Rutherford suspected that alpha particles were the same as fast-moving helium atoms, minus their electrons, but to prove it he needed to measure the particle's charge and mass, just as Thomson had done for the electron. One of Rutherford's recruits was Hans Geiger, who was working on an early version of what became known as the Geiger-Muller counter. The device, now used as a proportional counter, consisted of a 60-cm brass tube, evacuated to a low pressure. A thin wire ran down the center, with 1,000 V between it and the tube. This set up an electric field that was strongest near the wire. When alphas entered the tube from some small sample of radioactive material, they collided with atoms in the rarefied gas in the tube, knocking out electrons from atoms, leaving positively charged ions that were drawn towards the tube. The ions are accelerated by the electric field and collide with more gas particles, creating more ions, and so on. Free electrons would also move to the positively charged anode, the wire. The cascade effect meant that a single alpha particle could produce thousands of ions and free electrons that were drawn to the electrodes, creating a measurable voltage spike along the wire that could be transformed into a signal such as an audible click (Fig. 3.2).

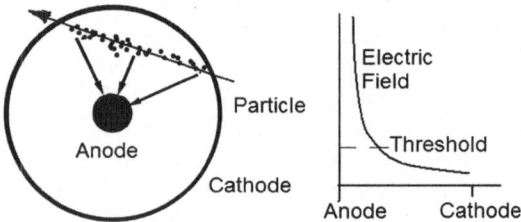

Fig. 3.2 Geiger counter basics (In a modified detector called the proportional counter, the electric field is shaped as shown on the right. In regions away from the central wire, where the electric field is below the threshold, the electrons from ionization along the particle path take time. Most of the avalanches happen closer to the anode. So the delay gives the location of the track)

Calculations using the (Geiger) counter implied that the alpha particle carried twice the charge of a hydrogen ion, which was consistent with the hypothesis that they were helium ions. To confirm the result, Rutherford's team used a second kind of detector, known as a scintillation screen, which had been invented by William Crookes (who also invented the Crookes tube which formed the basis for many experiments including the X-ray experiments). One day in 1903, Crookes had mislaid a small speck of radium that he was working with. He knew that radiation caused zinc sulfide to fluoresce, so, to find this piece of radium, he took a piece of zinc sulfide and moved it around his work area until it began to glow. (This, so-called scintillator is one of the first particle detectors invented. When a particle hits the scintillator, a trail of ionized atoms is created. The free electrons rapidly acquire energy from transfer processes and when these electrons lose their energy and recombine with the ionized atoms, a burst of light is released. The burst of light is brief and can be counted, to count the incident particle. The intensity of light is a measure of the particle's energy.) When Crookes took a closer look at the fluorescence through a magnifying glass, instead of a steady glow, he saw that the substance was giving off a marvelous display of individual flashes that seemed to shoot out at random. He deduced that the separate sparkles were produced by individual particles.

Crookes turned his discovery into a device called a Spinthariscope, from the Greek *spintharis* for spark, which consisted of a tube with a zinc sulfide screen at one end, a viewing lens at the other, and a miniscule speck of radium salt near the screen. The Spinthariscope was, for a while, a popular amusement among the Victorian upper classes, who weren't worried about staring at a radioactive material. For Rutherford, this proved that the radiation was emitted as individual particles and the scintillation screen enabled him to individually count the alpha particles as they were emitted from a thin sliver of radium, and confirm the results of the proportional counter. Finally, in 1908, the team proved that alphas were helium ions by collecting some in a tube, allowing electrons to join them to produce complete atoms, and showing that the resulting chemical was helium gas. Rutherford had therefore succeeded in determining what alphas were made of. Soon, he put this method right to work, as a tool to probe the interior of the atom.

While working at McGill, Rutherford had found that a beam of alpha particles, although it carried a charge, was not easily deflected using magnetic or electric fields. This implied that the particles had a high energy. On the other hand, if they passed through an extremely thin sheet of mica crystal (less than a thousandth of a millimeter thick) before reaching a photographic plate, they left a fuzzy image, which meant that the mica was able to scatter them from their path. It seemed that the forces in the mica were far stronger than the forces that could be imposed externally using magnets or electric fields.

Structure of Atom

In 1909, Rutherford returned to the question of alpha scattering. He asked Geiger and a student called Ernest Marsden to repeat the experiment, this time using a thin sheet of gold foil instead of mica, and a scintillation screen as detector. And rather than just placing the scintillation screen behind the foil, they tried placing it to the side, to see if the forces were so great that they could create really large deflections.

Counting the scintillations was hard work. Marsden and Geiger had to spend many hours in a darkened room, taking turns every minute or so to peer at the scintillation screen through a microscope and record the flashes. Rutherford was very appreciative of Geiger's work saying "Geiger is a good man and works like a slave.... is a demon at the work and could count at intervals for a whole night without disturbing his equanimity." From hundreds of thousands of observations, they were amazed to find that, while most of the alpha particles passed right through the gold foil, about one in 8,000 would bounce straight back off the foil. As Rutherford later put it, "It was as if you fired a 15-in. artillery shell at a piece of tissue paper and it came back and hit you."

The understanding of this comes from a somewhat straightforward calculation of the alpha deflection F from a plum pudding model. Since atom size (r_0) is about 10^{-10} m and the maximum force experienced by the incoming alpha particle, from the charge $79e$ of the gold nucleus (where e is the electron charge $= 1.6 \times 10^{-19}$ Coulombs), (if) distributed over this sphere of atom size as it passes the atom, would be (using Coulomb's Law of electrostatic forces)

$$F = \frac{q_t q_p}{4\pi\varepsilon_0 r_0^2} = \frac{(79e)(2e)}{4\pi\varepsilon_0 r_0^2} = \frac{158 \times (1.6 \times 10^{-19})^2}{4\pi(8.8542 \times 10^{-12}) \times (10^{-10})^2} \sim 3.6 \times 10^{-6}\,\mathrm{N},$$

(q_t is the charge on the target-gold atom(actually nucleus) $= 79e$, q_p is the charge of the incoming particle (alpha) $= 2e$). The force due to electrons is not counted, since electrons are tiny and we can assume that the alpha particle is not much affected by the electrons in the atom on average – gaining some energy from the electrons as they approach the electron and then quickly losing the same as they move away. The 5 MeV alpha particle with a mass $= 6.7 \times 10^{-27}$ kg would travel at a velocity (v) of about 1.5×10^7 m/s and therefore stay in the pudding sphere during the round trip (distance traveled is 2×10^{-10} m) for a duration t of about 1.4×10^{-17} s. So the impulse of the pudding positive charge would be

$$Ft = 3.6 \times 10^{-6} \times 1.4 \times 10^{-17} \sim 5.5 \times 10^{-23}\,\mathrm{N\,s}.$$

This is also the transverse momentum p that the alpha would gain, so that the transverse velocity

$$\delta v = p/m = Ft/m = 5.5 \times 10^{-23}/6.7 \times 10^{-27} \sim 8,200\,\text{m/s}.$$

This transverse velocity would give a deflection of the alpha ray by

$$\delta v/v = 8,200/1.5 \times 10^7 \sim 5 \times 10^{-4}\,\text{rad},$$

much less than a degree. But, to every one's surprise, the observed deflection in the experiment was sometimes 90° – thousands of times larger! The large deflection was only possible if the positive charge was very concentrated and the alphas could approach this concentrated charge. To be specific, this is possible only if the deflection was caused within by a positive charged sphere of radius $\sim 10^{-15}$ m. (Note above that the force F is inversely proportional to the square of the distance of closest approach and the transverse impulse and therefore the deflection is inversely proportional to this distance.)

The implication seemed to be that matter was mostly empty space, with central concentrations of positively charged material that repelled the alphas and therefore the alpha particle penetrated quite deep into the atom, over 10,000 times closer than previously assumed, before being scattered. This was an astonishing idea – it was like saying that even solid gold had no substance, but it was only an illusion created by a web of interacting forces (Aristotle would have protested vehemently). It was also completely inconsistent with the prevailing plum-pudding model of the atom, which assumed that charge was more or less uniformly distributed in space.

It took Rutherford a while to work out the details, but in 1911 he proposed a new model for the atom, in which a cloud of negatively charged electrons circle positively charged nucleus like planets around a sun. In place of the force of gravity, the attractive Coulomb force (see equation above) between opposite charges held the atom together, with electric charge replacing mass and the dielectric constant replacing the gravitational constant. Actually, this orbital model is an older model proposed by Hantaro Nagaoka as early as 1904. But Rutherford's experiment showed that the positive charge was concentrated in a very small core at the center of the atom. Alpha particles brushed easily past the extremely light electrons, but if one, by chance, was headed close to the far-heavier nucleus, it would be deflected by the nuclear charge. (Rutherford even simulated the process by swinging an electromagnet, pendulum fashion, from a thirty-foot wire, and directing it towards a second electromagnet on a bench oriented in such a way that it repelled the first.) With this close approach, the positively charged alpha particle would experience a large repulsive force and its path would be deflected by a large angle. Since he knew the charge and energy of the alpha particles, he could deduce from the scattering mechanics that the radius of the nucleus was about 10^{-15} m, or one hundred-thousandth of the atomic radius.

In a separate discovery, Rutherford found that when nitrogen gas was bombarded with alpha particles, the nitrogen atoms gave up a hydrogen atom stripped of its electron, which he called a proton. Just as Thomson had managed to strip the electron from the atom and determine its properties, so Rutherford had done the same for its positively charged counterpart. The proton gave the nucleus its positive charge. The loss of a proton in the nitrogen atom turned it into oxygen – another example of transmutation.

This model of the atom – with the lightweight, negatively charged electrons rotating around a positively charged core – was intuitive and easy to understand. However it had a problem, just like Nagaoka's model had, which Rutherford was himself aware of. In the solar system, the attractive force between the bodies is balanced by the centrifugal force of the planetary motion. In Rutherford's model of the atom, the electrons would have a similar balance of forces where the gravitational force is replaced by the electric Coulomb force. However, an orbiting electron is constantly accelerating in one direction and decelerating in the other and, according to Maxwell's equations, such electrons should radiate away their energy, slow down, and smash into the nucleus – all within less than a billionth of a second. So what was keeping the electrons in their place going around this new packet of new particles called protons? The answer to this mystery came from another branch of physics that was exploding into public consciousness in the early decades of the twentieth century: quantum mechanics.

The Quantum Jump

For a few brief years, around 1900, physicists had what they believed was a consistent view of the world, where some things like atoms were made of particles, and other things like electromagnetic radiation were waves. Newton had believed that light was just another form of particle, but since the early 1800s it had been known that light showed wave-like properties such as diffraction and interference. The cathode rays had been a matter of debate for some decades, but that ended when Thomson conclusively showed that they were made up of discrete particles with a measurable mass.

One puzzle, related to the light that was emitted from a so-called black body as it was heated, remained unsolved. According to classical theory, as a lump of iron is heated, the atoms would oscillate with increasing energy and give off electromagnetic radiation (including visible light) at a certain range of frequencies. However, the classical theory on this topic was a poor match for experimental observations – there was no way to tweak the classical model to make it work. The classical model implies that the body would radiate a continuous spectrum (all wavelengths) and this gives an absurd result (called ultraviolet catastrophe) that all its thermal energy is radiated away in an infinitesimally small time. In 1900 the German physicist Max Planck discovered, as a kind of mathematical trick, that the model could be made consistent with observations if it was assumed that the energy was not emitted in a continuous spectrum, but was released in a large number of finite, but very small, packets. The size of the packets was determined by a number that he called h, now known as Planck's constant $= 6.62606957 \times 10^{-34}$ Joule-second. The mathematics of quantized blackbody radiation is then bound by statistical mechanics and the derivation of radiation law is made more complex. This seemingly simple concept of quantized energy had serious implications (not realized then by Planck) and laid the foundation for the field of quantum mechanics, quantized fields and energies. Esoteric concepts of probabilities rather than determinism would prevail and uncertainties in time, space, momentum, energy, and reality itself and aspects of entanglement of these properties of particles would arise from it. The field of physics would be totally revolutionized.

A second puzzle was known as the photoelectric effect. Physicists had known for some time that light, when shone on a charged metal surface like the cathode in a cathode ray tube, could help liberate electrons to produce a spark. In 1902, Philipp Lenard found that the effect did not depend on the intensity of the light, but only the color, or more specifically the frequency. High frequency, blue light created a bigger spark than low-frequency red light.

In 1905, the young Albert Einstein, then working as a patent clerk, proposed a simple solution to both these problems. As Planck proposed, the blackbody effect could be explained if atoms were only allowed to emit or absorb radiation in discrete units of magnitude hv, where h was Planck's constant, and v was the frequency of the radiation. Similarly, Einstein argued that a beam of light of frequency v consisted of discrete parcels, each with energy hv. The particles were later given the name photons (in 1926 by Gilbert Lewis). The theory explained why the photoelectric effect worked only below a certain wavelength of the light, whatever the intensity. A high-frequency, high energy (small wavelength) photon could knock an electron out of its position in the atom, like a karate expert shattering a brick with a single blow. A large number of low-frequency photons had the same effect as a weak person trying to break the brick: what counts is the power of the punch, not the quantity. In the same year, Einstein further blurred the definition of matter by showing that energy and mass could be converted into one another, according to the equation $E = mc^2$.

Unlike Planck, Einstein, somewhat like Newton, saw the photons as real particle-like entities, rather than a mathematical abstraction. As he wrote, "Energy, during the propagation of a ray of light, is not continuously distributed over steadily increasing spaces, but it consists of a finite number of energy quanta localised at points in space, moving without dividing and capable of being absorbed or generated only as entities." Of course, not all of his colleagues were convinced. The American physicist Robert Millikan also spent 10 years trying to disprove the quantum theory, through careful study of the photoelectric effect, but only succeeded in confirming the theory, and obtaining an accurate measurement of Planck's constant. (As compensation, he was awarded a Nobel prize for his efforts, a year after Einstein's).

In 1912, armed with this new concept of quantized emission, the Danish physicist Niels Bohr visited Manchester, where he worked with Rutherford on a new model of the atom that combined elements of the classical and quantum approaches. In the Rutherford–Bohr model (Fig. 3.3), electrons circled around the nucleus in orbits as before, but since they cannot lose energy continuously and can only emit a discrete quantum of radiation, they change orbit (jump) to one with less

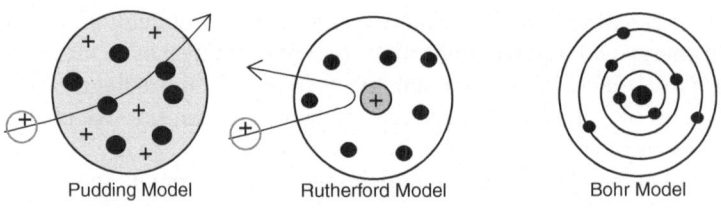

Pudding Model   Rutherford Model   Bohr Model

Fig. 3.3 Bohr-Rutherford Model (*right*) and the earlier models for the atom

energy. The reason, all the electrons do not just jump off to lower orbits and emit photons until they smashed into the nucleus, was because each orbit only had room for a fixed number of electrons, and those in the innermost orbit were prohibited from moving in.

The model was purely a work of imagination, but its advantage was that it correctly predicted how substances could emit radiation. It had long been known that when a particular element is heated, it gives off a unique spectrum of radiation, with peaks at particular frequencies (line spectra). In the Rutherford–Bohr model, these characteristic frequencies corresponded to the difference in energy between adjacent electron orbits. The model correctly predicted the wavelengths of the Fraunhofer lines (hydrogen spectral lines in solar spectrum).

Development of the quantum theory occupied top theoretical physicists for the first few decades of the twentieth century – and completely overthrew our notions of reality. In 1924, Louis de Broglie submitted a doctoral thesis, to the Sorbonne Conference in Paris, which provocatively suggested that if light could behave like a particle, then a particle could be viewed as a wave. The wavelength, he computed, should be equal to Planck's constant divided by the momentum of the particle. If the particle was sufficiently small, then the wavelength would be large relative to the size of the particle and vice versa. Furthermore, de Broglie proposed that an electron's orbit around the nucleus had to correspond to an integer number of wavelengths. Later proper quantum mechanical descriptions would show that the orbital electrons exist in an orbit in the form of a standing wave, like a plucked guitar string with a wavelength $\lambda = h/p$ where p is the electron momentum, that can be fitted in the available length, say the circumference of the orbit. Therefore, an electron can have only discrete momentum or wavelength in a given orbit. The different orbits, with distinct energy levels, corresponded to the different possible harmonics and each harmonic resonant mode could only have a specific energy. This latter point fits the observed atomic orbital stability and the quantized energy emission, since this correspondence to a resonant orbit would preclude emission of arbitrary amounts of energy. A wave is described by an equation that relates the time dependence of deflection of a particle in a medium (say water) or the amplitude of a wave at one location to the variation of these deflections/amplitudes as a function of location. This wave equation then became the basis for the famous Schrödinger equation, which could be used to derive properties of atoms and particles, their energy states, and interaction with electric and magnetic fields.

The behavior of the particles as waves is also an observed fact and is used, for example, in electron microscopes to image molecules and surface structures.

Experimental evidence that electrons had a wave-like quality was supplied independently by Clinton Davisson and Lester Germer in the USA, and by George Thomson (son of J.J.) in Scotland. Thomson and Davisson were later awarded the Nobel prize, which resulted in the oft-remarked irony that J.J. Thomson won a Nobel prize for proving that the electron was a particle, while George won it for proving that it was a wave. The mathematics was daunting, but quantum theory was finally put on a firm theoretical footing by physicists such as Edwin Schrödinger, Werner Heisenberg, and Paul Dirac. The dual wave-particle nature of matter was summarized by Heisenberg's uncertainty principle, which stated that a particle's position and momentum could never be perfectly known at the same time. The more precisely you know the position, the less sure you can be of the momentum, and vice versa. This helped make sense of Bohr's assertion that electrons in the innermost orbit could never be drawn into the nucleus. If that were to happen, the electron's position would be known to the size of the nucleus, and its momentum would also be known to be near zero, thus violating the uncertainty principle. While the wave nature of particles is clearly seen in the governing equations and is available for statistical interpretations, its conceptual meaning has remained a matter of debate. The description of particles also as waves gave Paul Dirac's proof for the existence of antimatter (see Chap. 6). Dirac had predicted the existence of antimatter and creation of matter out of pure energy and provided creative fuel for space ships in Star Trek.

Discovery of the Neutron

In 1930s, the call for a program of accelerator research was speculative – and there were more immediate things to worry about. Rutherford, who had returned to Cambridge to head the Cavendish laboratory, having named the particle in the nucleus of the hydrogen as proton, had also hypothesized the existence of a neutral particle in the nucleus, which he called the neutron. The reason was that the atomic weight of many elements could not be explained from the protons alone. For example, hydrogen consisted of an electron and a proton, so had an atomic weight of 1; while the next heavier element helium had two electrons, so for it to be electrically neutral it needed to have two protons. But its atomic weight was not 2, but 4 – so something neutral had to be providing the missing mass.

Rutherford initially believed that the neutron was a doublet consisting of a proton and electron bound together. This would explain how high-energy electrons appeared to be emitted from the nucleus in beta radiation. But when he tried to explain the mechanics to an audience in 1927, few were convinced. As a student wrote, "The crowd fairly howled. I think Rutherford came nearer to losing his nerve than he ever did before." Rutherford needed experimental results to prove the existence of the neutron. But if the neutron had no charge, it would be unaffected by electric or magnetic fields – so how could he possibly detect it?

A hint came in August 1930, when Walter Bothe and his assistant Herbert Becker published results from an experiment, in which they had used alpha particles from the radioactive material polonium to bombard light metals such as beryllium. Using a Geiger counter, they found that the target material produced some powerful rays, which they believed might be gamma rays. Frédéric and Irène Joliot-Curie (Irène was the daughter of Pierre and Marie Curie) followed up on their work, and found that the rays from the beryllium could cause protons – i.e., hydrogen ions – to be ejected from paraffin (which is rich in hydrogen). The effect was similar to the photoelectric effect, where ordinary light ejects electrons from a surface, with the difference that outgoing protons are about 2,000 times more massive than electrons – so the radiation had to be tremendously powerful.

When James Chadwick at Cavendish heard about these results, he immediately connected them with Rutherford's neutron. Gamma rays were powerful, but they were high-energy photons and so had no mass. A neutron, in contrast, had the same mass as a proton and so could easily knock the proton from its perch. Chadwick replicated the experiment by building a kind of ray gun. He managed to obtain some of the expensive polonium, from Kelly Hospital in Baltimore, in the form of a penny-sized disc shaped thin film. He then simply placed the disc next to another disc of beryllium inside an evacuated tube. The polonium emitted its alpha radiation, which hit the beryllium and produced the mysterious rays, mostly in the direction away from the polonium. It was like a machine gun that (almost) never ran out of bullets (Fig. 3.4).

Chadwick first aimed his device at a detector and recorded the radiation emitted from a window at the end of the tube. He then placed a 2-mm sheet of paraffin between the gun and the detector, and confirmed that a spray of protons was produced. To measure their energy, he tested how far they could penetrate through thin sheets of aluminum foil. He calculated that the protons had an energy of 5.7 MeV. This was

Fig. 3.4 James Chadwick's neutron experimental apparatus. Image 10313943, Science Museum, Science and Society Picture Library

far more energy than could be produced by a single photon. However, it was consistent with the idea of a heavy neutral particle that could penetrate the atom without being deflected by electric charge. As Chadwick later wrote, "If we suppose that the radiation is not a quantum radiation, but consists of particles of mass very nearly equal to that of the proton, all the difficulties connected with the collisions disappear."

The details of the reaction are as follows: the polonium emits an alpha particle. This collides with a beryllium nucleus, of atomic weight 9, to create a carbon atom of weight 12, plus a free neutron. Chadwick didn't know it at the time, but his simple gun was the first step on the road to a much more powerful weapon: the atomic bomb. In 1938, the German scientists Otto Hahn and Fritz Strassmann in Germany found that if neutrons were directed at uranium, they split the nucleus in two. This released energy, which in turn produced more neutrons, and so on. If enough material was present, the result was a runaway chain reaction. Chadwick later wrote that, on realizing the inevitability of the nuclear bomb, "I had then to start taking sleeping pills. It was the only remedy."

Democritus's atom was about to go big time. The first three decades of the twentieth century had already changed human understanding of the nature of matter and forces, beyond imagination. The atom having been described, the nucleus was ready to reveal its secrets. The nuclear era had dawned. Nuclear bombs and reactors, with their economic, social, and political implications, were inevitable. The two giant leaps, the theory of relativity and quantum mechanics had changed the landscape of physics and science, in general. Particle streams from accelerators working in natures' basement would be inadequate to probe matter further.

While theorists were tinkering with their equations, and attempting to penetrate the atom using the finely honed tools of their intellects, experimentalists like Rutherford were wondering if they could find their way in using a more direct means. The problem was that particles from radioactive sources had a fixed amount of energy. In principle, there was no reason why particles could not be accelerated to greater energies using the new technologies that were being developed for high-voltage electrical transmission. The result would be "a copious supply of atoms and electrons which have an individual energy far transcending that of the alpha and beta particles from radioactive bodies," as Rutherford put it in his 1927 presidential address to the Royal Society.

Invisible Rain, Cosmic Rays

"Round about the accredited and orderly facts of every science there ever floats a sort of dust-cloud of exceptional observations, of occurrences minute and irregular and seldom met with, which it always proves more easy to ignore than to attend to… Anyone will renovate his science who will steadily look after the irregular phenomena, and when science is renewed, its new formulas often have more of the voice of the exceptions in them than of what were supposed to be the rules." so said,

William James, eminent psychologist. A year after the demise of this great mind, Victor Hess would prove once again, how right he was.

Like on six previous occasions, on a balmy dawn on Wednesday in August 1912, the orange and black, 12 stories tall balloon was inflated. The "Bohmen" rose majestically to its full height as the members of the Austrian Aeroclub inflated the balloon to its full load capacity of 2 ton. Captain Wolfgang Hoffory, a veteran pilot and Dr. Victor Hess, a 29-year-old physics teacher at the Viennese Academy of Sciences swung aboard and adjusted the ballasts. Once the ropes were untied, the balloon rose over the meadows in Aussig, Bohemia and eclipsed the sun. As it did, Dr. Hess opened up his wares, exposing three 2 liter size electroscope chambers that measure level of radiation, taking periodic readings with great attention and accuracy, using a microscope to view the electroscope foils through a small window in the chamber. The itinerant balloon fliers kept rising and buffeted by the freezing 30 km winds, the Bohmen ascended to a terrifying height of 5,350 m (about 18,000 ft). Dr. Hess continued his measurements undaunted as Hoffory piloted the balloon adroitly. The descent took 2 h for a total flight duration of over 6 h and through it all, Dr. Hess kept his eye on the electroscope and took diligent readings, little knowing that all these measurements would be another giant leap for physics. Like on April 12, 1912, when he made a balloon flight during solar eclipse, this too yielded a wealth of data (Fig. 3.5) [John Kraus, Cosmic Search, Vol.2, No.1, p.20 (1980)].

Fig. 3.5 Victor Hess readying for his balloon flight, "This month in Physics", APS Physics news, Vo. 19, No.4 (April 2010), Ed. Alan Chodos

In 1912, the subject of radioactivity was, well...hot! So, many scientists had electroscopes, a device invented by the physician William Gilbert in 1600. It could measure charge and was later adapted as a gold leaf electroscope by Abraham Bennet in 1787. As the metal at the top of a sealed jar was charged by touching an electrically charged rod, two gold leaves connected to it inside would too charge to the same sign, repel each other, and flare out. For a long time, it was a puzzle that the electroscopes would slowly lose charge and the leaves would droop. Then they figured out that radioactive radiation penetrated the jar and ionized the air inside

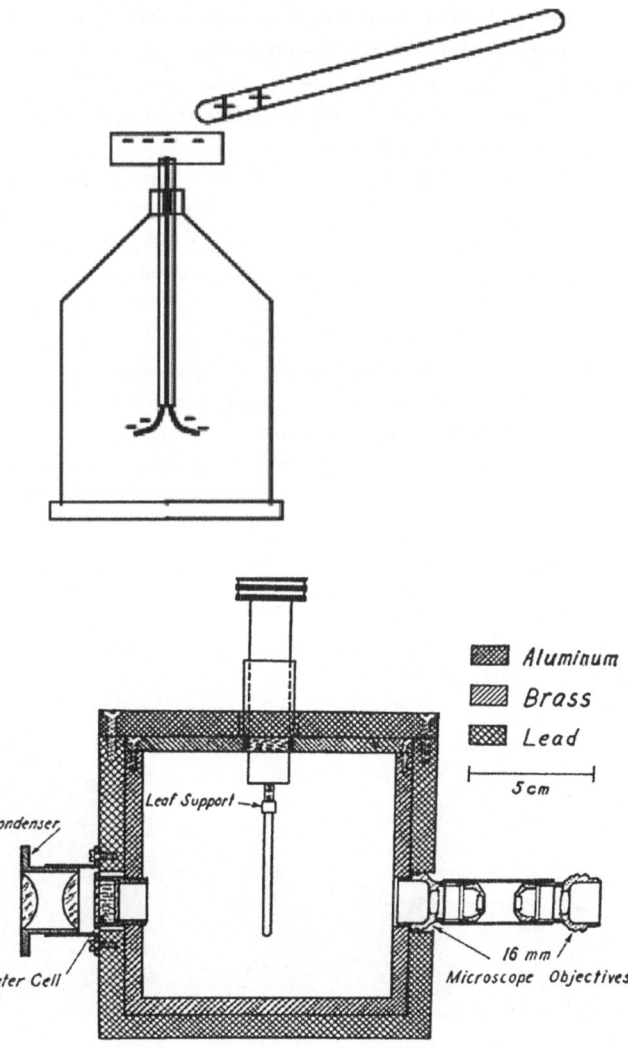

Fig. 3.6 *Top* – Schematic of a gold leaf electroscope, *Bottom* – U.S. National Bureau of Standards electroscope for measuring radiation (NIST, USA). Dorsey, N.E., *Physics of Radioactivity*, 223 pages, Williams and Wilkins Co., Baltimore, 1921

and the free ions and electrons neutralized the charge on the leaves. The rate at which the leaves dropped is proportional to the level of radioactivity. This then became an instrument for measuring ionizing radiation (radioactivity). The electroscope with solid-walled chambers would prevent any stray radiation, permitting only the desired ones allowed to enter through a window and the rate of closing the leaves would be measured through a microscope with a graduated scale, to determine the intensity of radiation (Fig. 3.6).

While there was firm knowledge that radioactivity caused the loss of charge, the leaves dropped even without radioactive materials and this loss was found everywhere. So there must be another source of radiation. Victor Hess wanted to see if all radioactivity was terrestrial. If it were, then the radioactivity should decrease as we go above the ground, which was observed to be true over a certain height. After Hess came down from his flight and plotted his data on radioactivity as a function of height he saw the trend shown in Fig. 3.7.

As Hofforoy brought the balloon up over about 1 km, the level of radioactivity started to increase and then kept increasing. Hess concluded that there were radiation sources in the sky. In his report in November 1912, he stated, "The results of the observations indicate that rays of very great penetrating power are entering our atmosphere from above." On June 28th, 1914, German scientist Werner Kohlhorster took a balloon flight to the phenomenal height of 9,300 m (about 33,000 ft). His measurements corroborated Hess's conclusion. As he landed, Archduke Francis Ferdinand, the Austrian-Hungarian heir-apparent, had been assassinated and the world was plunged into the first world war.

Hess's April measurements confirmed that while Sun was a source, there were also other sources, which remained even during a solar eclipse. Hess thought the

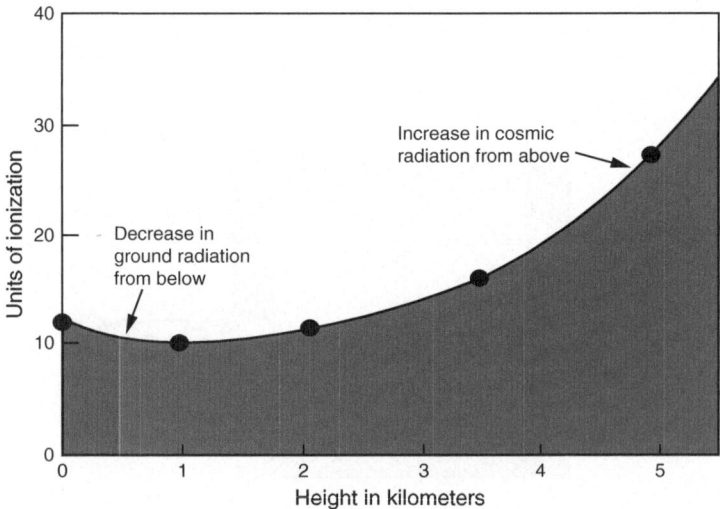

Fig. 3.7 Variation of radioactivity as a function of balloon height, as seen by ionization events counted by the ionization events in the electroscope

Fig. 3.8 Terrestrial radioactivity and cosmic showers were a gift to the physicists and a window into particle types and properties

source was "atmospheric electricity." Robert Millikan of Caltech proved that these were rays of extraterrestrial origin and between 1934 and 1937, Bruno Rossi and Pierre Auger showed that two detectors spaced far apart detected radioactivity simultaneously, which led to the conclusion that a primary ray entering the atmosphere was causing a broad secondary shower of multiple particles that arrive simultaneously. In 1948, Melvin Gottlieb and James Van Allen showed that the main components of cosmic rays were protons and alpha particles. We now believe that while lower energy rays are born in the sun, the high energy cosmic rays are born out in space, most likely in exploding stars where large accelerating fields are created by swirling fluids in megaGauss magnetic fields, accelerating particles to energies of billions of electron volt (1 electron volt = 1eV = 1.6×10^{-19} J = energy gained by an electron in an accelerating potential of 1 volt).

Cosmic rays provided the fillip for new physics discoveries, by unfailingly ladling out, then unrealized, continuous stream of high energy particles (Fig. 3.8). Even today, cosmic rays are consistently used in physics experiments both for basic physics and for astronomical observations. But true to human nature, human endeavor would seek to create its own supply of such tools of research and control the supply of energetic particles.

Chapter 4
Cracking the Nucleus

Once upon a time, the son of a great chief suddenly saw his friend disappear after he made fun of the moon. In a listless state, he started shooting arrows at the moon. The arrow disappeared, and he was encouraged. He thought, "Now I am going to shoot that star next to the moon." In that spot was a large and very bright star. He shot arrows at this star, and sure enough, the star darkened. After some time, he saw something hanging down close to him, and, his next arrow stuck to it. The next did likewise, and at last, a chain of arrows reached him...Now the youth felt badly for the loss of his friend and, lying down under the arrow chain, he went to sleep. After a while he awoke and looked up. Instead of the arrows, there was a long ladder reaching right down to him. He climbed up the ladder to look for his friend in the sky.

Adapted from Tilingit Indian mythical tale (S. Thompson, Tales of the North American Indians (Bloomington, 1966), p. 132).

The year is 1932. It is, of course, the momentous year in which James Chadwick identifies the neutron as a constituent of the nucleus. The period is also famously marked. Aldous Huxley's novel "The Brave New World" is published and will make the world think about the future for generations to come. In India, 2 years back, Mahatma Gandhi had planted seeds for nonviolent struggle by going on the "Dandi March," to make his own salt at the beach in an act of civil disobedience. Jack Benny goes on television to entertain audiences for decades to come and sets a standard for comedy. But this was also the year in which the nucleus gave up the first of its inner secrets to the experimentalists. Quantum mechanics had been discovered only around 1925. This theory coming at the heels of another spectacular theory, the Theory of Relativity, had marked the genius of the theoretical physicists. But, the experimentalists were as deeply invested in admiring and understanding the mysteries of the Universe as the theoretical physicists were and this was to be their decade.

If you could journey into the past to around 1930, you would notice that physicists already know that the elements are distinguished by the amount (number) of nuclear charge, which is equal to the number of electrons in the atom. They also

R. Jayakumar, *Particle Accelerators, Colliders, and the Story of High Energy Physics*, DOI 10.1007/978-3-642-22064-7_4, © Springer-Verlag Berlin Heidelberg 2012

observed that isotopes, that is, elements with the same number of electrons have the same electronic structure and have nearly identical chemical properties (except for the kinetic isotope effect which slowed down reaction rates). They knew that the charge was given by the number of protons equal to the number of electrons outside the nucleus. But, they don't quite know what to do with the number of neutrons, which seemingly added only mass and created a zoo of chemically identical isotopes. They were very dissatisfied with the fact that they have not cracked open the nuclear nut to see inside the nucleus, and watch the neutrons and protons spilling out. To quote Rutherford – like children, they needed to take the watch to pieces to see how it works.

One needed a particle bullet to penetrate the nucleus. It was logical that one would choose a positively charged particle such as the proton for the bullet, so that the Coulomb force of the electrons that are outside the nucleus would attract it. But, as the particle penetrates the electron cloud and approaches the point-like positively charged nucleus, the proton would be repelled by the Coulomb force, which increases as the inverse of the square of the decreasing distance of the approaching proton. The electric repelling force of the nucleus presents a steep potential hill to the proton, and the particle needs a high energy to overcome this. While at that time physicists did not know of the strong force inside the nucleus, they knew something bound the protons together and so all they needed was to get the incoming particle across this repelling shell and the particle would act as a nuclear probe. Conventional wisdom said that with the small size of the nucleus and its powerful charge, the probing proton needed to have the energy of several million electron volts. For example, the electron in a hydrogen atom is at a distance nominally at about 10^{-10} m and is bound to the nucleus with energy of the order of 10 eV and requires that much work to free it. But a nucleus is about 100,000 times smaller and therefore, the potential (force times the distance), which increases inversely with decreasing distance, rises to over a million electron volts and so the particle energy has to exceed this. Though such an energy was available from radioactive sources, the beam of particles was too feeble for carrying out a quantitative experiment and energies of few MeV from particle sources was a tall order in that year.

The Russian born George Gamow came as a white knight to the rescue, riding the white horse of quantum mechanics. Quantum mechanics had been invented only a few years earlier and Gamow was one of the first to understand it. In 1928, Gamow had solved the problem of alpha decay of nuclei, where the alpha particle overcame the attractive force of the nucleus to emerge outside with sufficient energy to escape it. So, he proposed that if a particle can escape the potential from inside, a particle can cross it from the outside. Quantum mechanics was this new found quizzical field (see Chap. 3) in which physicists found that reality at a microscopic level was much more fuzzy than the well-determined and well-calculable classical mechanics expounded by Newton. While motorcycles and skateboards need the calculated approach speeds to climb a slope, at atomic and nuclear sizes, particles can "tunnel" (symbolically) through this hill and find themselves on the other side, without really climbing the top (Fig. 4.1).

Fig. 4.1 Gamow explained that there is a probability that a particle does not have to tally climb the wall, but can sneak through the wall (symbolically and semantically a tunneling process, but in quantum mechanical description, is really a probability of finding it behind the wall)

The origin of this tunneling arises from two aspects of the nature of matter, described by quantum mechanics. (1) particle energies are quantized: energies exist only in discrete steps, proportional to the Planck's Constant h. Classical mechanics, of course, corresponds to this constant being zero. (2) Particles also behave (travel) as waves, the wavelength being $\lambda = h/p$, where p is the particle momentum. In free space, these waves are unattenuated, but when they encounter a barrier they are gradually attenuated (amplitude decreases slowly), rather than stopping abruptly as in classical mechanics. These two features are encapsulated in the Schröedinger equation (first published in 1926 and heartily endorsed by Albert Einstein), which describes the temporal and spatial behavior of a particle wave function using a wave function $\psi(x, y, z, t)$, which is a description of the state of the particle, in terms of its position coordinates x, y, and z and time t.

A wave function of a particle is a quantum mechanical description and has no direct physical meaning. It is a complex quantity with real and imaginary parts giving amplitude and phase of the wave. This is not unlike the description of light (electromagnetic) or sound waves. In quantum mechanics, however, a particles state is a composite of all possible states, some with more and some with less probability (in the statistical interpretation of the wave theory). Thanks to the description developed by Max Born, the square of the modulus of the wave function $|\psi|^2$ does have a physical meaning and is the probability (density) that the particle may be found in that state. Therefore, how it varies with time and space describes the probability of a particle being one place or the other at one time or another and with certain energy. In common applications, the wave function is "normalized" so that the sum of probabilities of all states is 1 (i.e., 100%).

The Schroedinger equation is then

$$i\hbar \frac{\partial \psi}{\partial t} = E\psi = -\frac{\hbar^2}{2m}\nabla^2\psi + V(x,y,z)\psi \tag{4.1}$$

where m is the mass of the particle moving through a potential V varying over space. (The $\frac{\partial}{\partial t}$ operation is a partial differential with respect to time t only. The del ∇^2 operator is the second order differential with respect to space coordinates, $= \frac{\partial^2}{\partial x^2} + \frac{\partial^2}{\partial y^2} + \frac{\partial^2}{\partial z^2}$, in the Cartesian coordinate system of x, y, z). This equation states that the time variation of the wave function is related to the spatial distribution of the wave function and therefore a particle with a velocity is like a wave that is traveling. These equations describe different states of the traveling particle and solutions to this equation predict the specific value of wave functions $\psi(t, x, y, z)$ at an instant and location.

The above equation can be rewritten as,

$$\nabla^2\psi = -\frac{2m}{\hbar^2}(E - V)\psi \tag{4.2}$$

if we consider only one spatial dimension x, then

$$\frac{d^2\psi}{dx^2} = -\frac{2m}{\hbar^2}(E - V)\psi \tag{4.3}$$

with solutions of the form (where, for illustration, the potential is assumed to be constant, that is, the hill is a square barrier, unlike in the figure)

$$\psi = Ae^{ikx} + Be^{-ikx}, \quad \text{where } k = \sqrt{\frac{2m(E - V)}{\hbar^2}}. \tag{4.4}$$

and $i = \sqrt{(-1)}$ For $E > V$ (energy of the particle greater than the barrier), the complex function

$$e^{\pm ikx} = \sin(kx) \pm i\cos(kx)$$

is a sinusoidal waveform representing a wave (particle) going in the $+x$ direction (forward) or in the $-x$ direction (reflected). Equation (4.4) can also be written as,

$$\psi = (A + B)\sin(kx) + i(A - B)\cos(kx) \tag{4.5}$$

$|\psi|^2 = 2(A^2 + B^2)$ is the probability of the state ψ (also interpreted as fraction of the particles with properties corresponding to that state).

But for $E < V$,

$$\psi = Fe^{-px} + Ge^{px}, \quad \text{where } p = \sqrt{\frac{2m(V - E)}{\hbar^2}}, \ k = ip \tag{4.6}$$

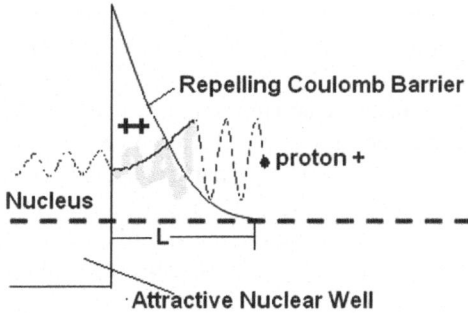

Fig. 4.2 The potential barrier due to surface charge on the nucleus, faced by the proton (wave) approaching a nucleus. The Coulomb barrier is faced by the proton, after it has passed through the attractive potential due to orbiting electrons outside the nucleus. Once the attenuated wave tunnels through the barrier, it was speculated that the wave would see an attractive potential (though not investigated then)

and both terms are real, but the second term with the coefficient G is zero, since otherwise the solution e^{px} continuously increases with x and the probability of finding the particle ($|\psi|^2$) would absurdly increase with distance. The decaying (attenuated) component (evanescent wave) with the amplitude F (see Fig. 4.2) is what we would find for our low energy intrepid proton inside this nuclear fort of a hill. This states that the wave function decays exponentially, as the wave enters the hillside and so does the probability of finding the particles beyond the hill (at twice the rate of the wave function).

For a mighty hill $V >> E$, the approximate transmission coefficient (the ratio of probability of finding the particle on the other side of the barrier to the probability of finding it just before the barrier) $= |F|^2/|A|^2$ is e^{-2pL}, where L is the barrier width. Now note that, to the experimenters delight, the probability of finding the particle inside this hill $|\psi|^2$, though decayed, is still nonzero. Though the potential is larger than the energy of the particle, the wave does not stop; it only attenuates to a lower amplitude. Summarizing, quantum mechanical statistics says that for a given particle energy, there is a nonzero probability that the particle can be found on the other side of a hill even with a hill potential larger than the particle energy. This result is distinctly different from Newtonian mechanics, which would forbid particles with $E < V$ entering the hillside. The Newtonian result can be obtained by setting h tending to zero; p then tends to infinity and e^{-2pL} tends to zero. The smaller $(V–E)$ is, (the smaller the hill or higher the energy), the smaller is p and higher is this probability of transmission. A smaller shell width L of barrier of the nucleus also increases the transmission coefficient. Since p increases with m, lighter particles have a higher probability of tunneling.

This description has a complementary (alternative) particle-based description in the Heisenberg school. One could say that there is a probability that the proton has more energy than we measure it to be. This comes out of the famous Heisenberg uncertainty principle. The readers might have come across this in pop culture (for example, Science Fiction books such as Gap cycle and TV series like Star Gate and

Futurama) in which the principle is invoked by saying that if you know one property very accurately, then you cannot know a complementary property accurately. (In fact, a version of the origin of the Big Bang and birth of alternate Universes invokes this very principle.) Since distance and momentum are complimentary quantities, as the distance to the hilltop decreases, the location of the particle becomes more accurate and the speed (momentum) of the particle has large uncertainty. The larger the momentum of the proton, the larger is the uncertainty and the probability that more protons would be found on the other side. (The principle of tunneling has widespread application and was used in inventing the tunnel diodes, a common feature in electronic circuits.)

This was great news for the experimentalists. If it was probable that low energy alpha particles can climb the hill and get out of the nucleus, some lower energy protons from outside the hill can climb into the nucleus. You need enough nuclear penetrations for obtaining good statistics for data. So, you needed a lot of protons to assault this fortress so that at least some of the warrior protons can get in. The lower the proton's energy, the more intense the proton beam would need to be. There was no radioactive source that could provide a high intensity of proton beams. The physics community had come to a momentous conclusion exemplified by Ernest Rutherford's earlier address at the British Royal Society [See The Fly in the Cathedral by Brian Cathcart, Farrar, Straus and Giroux, (2005)]. The burly and formidable Briton and Director of the famous Cambridge's Cavendish lab declared:

> It would be of great scientific value if it were possible in laboratory experiments to have a supply of electrons and atoms of matter in general, of which the individual energy of motion is greater even than that of the alpha particle. This would open up an extraordinary new field of investigation that could not fail to give us information of great value, not only in the constitution and stability of atomic nuclei but also in many other directions.

The conclusion was that experimenters could not be satisfied with the high energy particles provided by nature such as alpha particles coming from radioactive decay or cosmic ray particles, but needed accelerators to produce intense beams of high energy particles. This was a turning point in the history of physics, giving rise to the birth and the present era of accelerators. This would set a new paradigm in methods of experimental physics that is based on a global consensus on the needs of the science. This decision, to build large machines with a large budget, probably germinated today's Big Science Projects, such as the Space programs, the Hubble Telescope, Genomics, Deep Sea Drilling Project and of course, the large High Energy Physics and Nuclear Fusion experiments.

The Cockroft-Walton Generator: One of the First Accelerators

The particle accelerators considered at the time were DC high voltage sources. The sources would generate a high accelerating voltage at one electrode. The particle sources would be placed at the other electrode and accelerated by the applied

Fig. 4.3 Cockroft Walton
generator used as first stage of
accelerator (injector) in the
Alternating Gradient
Synchrotron in Brookhaven
National Laboratory
(Courtesy: Brookhaven
National Laboratory, NY,
Image No. CN7-1397-69)

voltage. The high energy particle would be extracted through a hole or a grid at the high voltage electrode. Working from an idea developed in 1919 by a Swiss physicist Henry Greinacher, J.D. Cockroft, and E.T.S. Walton, researchers at the Cavendish Laboratory in England had worked out a method to double and quadruple the initial voltage that is supplied at the input of a generator. The idea departed from the common use of direct current electricity. The CW (Cockroft–Walton) generator (Figs. 4.3 and 4.4) uses alternating current and a string of capacitor bridges. By appropriately switching the charging paths with a switch or a diode during the alternating cycle of the charging circuit, different capacitors get charged and build up voltage at the output. (Capacitors are devices commonly used for storing electric charge for short times in electric circuits and diodes are devices that permit current to flow in only one direction.) For a given voltage rating, the more the number of capacitor–diode stages, the higher would be the voltage. Also, a capacitor with a larger capacitance would store more charge, and therefore can supply more current for a given voltage. The Cockcroft–Walton accelerator achieved nearly 800 kV in 1932 starting from an input voltage of 200 kV. At this time, this was a huge energy and the scientists were eager to use this generator. The CW generator provided copious amounts of accelerated protons, electrons, and other charged particles and experimentalists with their big appetite were thrilled.

While the above figure gives an idealized description, the actual voltage obtained does not increase with number of stages linearly because of resistive losses in capacitors, switching losses (e.g., drop across the diodes), and energy drain by accelerating a beam of particles. This loss in voltage actually increases as the cube of number of stages. This nonideal effect can be reduced if the capacitance of the capacitors is large and the frequency of the input voltage is high. (But very high frequencies may again increase losses in components.)

A team headed by Rutherford himself shot protons accelerated with CW generator, on targets made of lithium. Initially, after looking unsuccessfully for gamma rays produced by nuclear interaction, they tried the new kind of particle detector

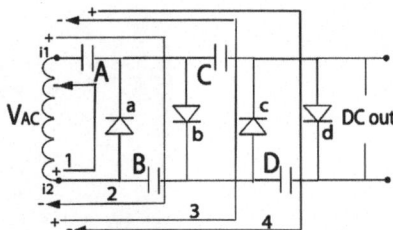

Fig. 4.4 Cockroft-Walton half cycle generator

In this figure, AC voltage with a peak value of V is supplied at the input ends i1 and i2, after 4 cycles of oscillation of the input voltage Vac $= V$ (the current flow direction is labeled with the cycle sequence number), a high D.C. output $= 4$ times the applied voltage is obtained

In cycle 1: i1 is negative, i2 is positive, diodes a and c are conducting; diodes b and d are not conducting. Capacitor A charges to the Input voltage V through diode a, there is no voltage to drive current through c

In cycle 2: i1 is positive, i2 is negative; diode b and d are conducting; diode a and c are not conducting

   Capacitor B charges to a voltage $=$ Input voltage V + Voltage on A (V) $= 2$ V through diode b; there is no voltage to drive current through d

   In cycle 3: i1 is negative, i2 is positive, diode a and c are conducting; diodes b and d are not conducting

   Voltage across diode a $=$ Input voltage V + Voltage across A $(-V) = 0$; hence no current through a. Capacitor C charges through c to a voltage input V + Voltage across B (2V) $= 3$V

   In cycle 4: i1 is positive, i2 is negative, diode b and d are conducting; diode a and c are not conducting. Voltage across b $=$ Input voltage V + Voltage across A (V) + Voltage across B $(-2$ V) $= 0$; hence no current through b; D charges through d to a Voltage $=$ Input Voltage V + Voltage across C (3 V) $= 4$ V (Top output terminal is negative)

made of zinc sulfide. When they started using the scintillators, to their great delight, they saw strong signals indicating bursts of alpha particle emission. The nuclear reaction that resulted is given by $Li_3^7 + H_1^1 = 2He_2^4$ that is emission of two alpha particles for one proton hitting the lithium nucleus. Cockroft and Walton reported the following in Nature, 129, 649 (1932),

> To throw light on the nature of these particles, experiments were made with a Shimizu expansion chamber, when a number of tracks resembling those of a-particles were observed and of range agreeing closely with that determined by the scintillations. It is estimated that at 250 kilovolts, one particle is produced for approximately 10^9 protons.
>
> The brightness of the scintillations and the density of the tracks observed in the expansion chamber suggest that the particles are normal α-particles. If this point of view turns out to be correct, it seems not unlikely that the lithium isotope of mass 7 occasionally captures a proton and the resulting nucleus of mass 8 breaks into two a-particles, each of mass four and each with an energy of about eight million electron volts. The evolution of energy on this view is about sixteen million electron volts per disintegration, agreeing approximately with that to be expected from the decrease of atomic mass involved in such a disintegration.

In an excellent demonstration of Gamow's theory, when they reduced the voltage to 150,000 V, they still got a signal (albeit reduced) of alpha particles.

There is a subtlety to this result. As noted before, multimillion electron volt alpha particles were available to experimentalists at the time from radioactive sources, but due to the nature of this quantum mechanical tunneling process, several hundred thousand eV protons can get closer to the nucleus than the multi-MeV alpha particles. (Note the exponential dependence of tunneling coefficient on the probe particle mass.) The success of this momentous experiment was because of the availability of a copious supply of protons from an ion source and accelerated through a CW generator which provided large enough number of nuclear interactions. The lithium experiment also demonstrated Einstein's theory on the equivalence of mass and energy, since the kinetic energy carried by the alphas resulted in a net mass reduction (resulting from change in nuclear binding energies).

While that year's discovery of neutrons by Chadwick was the most important discovery in physics, these results were also historic for two reasons. One, this was the first ever realization of what had always been dreamed of by alchemists – a man-made reaction to convert one element into another. In this case, the reaction had converted lithium into helium atoms. Second, this was the first "nuclear" experiment with the first high-energy accelerator. David had downed Goliath with a puny accelerated piece of stone.

Cockroft and Walton's experiment was the first successful attempt to split the atom with the release of alpha particles with millions of electron volts of energy. This was a particular feather in the cap of these experimenters because they felt they beat the American team, headed by the future great E.O. (Ernest Orlando Lawrence), who was actually still behind on his extremely promising concept of a "Cyclotron." As always, the public and the press leaped ahead of the scientific results, and in a then prevailing atmosphere of fear of war, the CW machine and the results were seen as harbingers of splitting of nuclei in chain reactions (sustained nuclear fission) and atom bombs, although these had not yet been conceptualized. There is another reason to celebrate this experiment. This was one of the first collaboration between physicists and engineers in what would become a thriving industry, bringing technological spin-offs of such collaboration into the consumer industry. Once again, this demonstrates the role fundamental research has played in the daily lives of people.

Over the years, the Cockroft–Walton generator was perfected using such a collaboration, with ever improving and more compact components. Even today, the CW generator is used to accelerate particles and inject them into bigger accelerators (Fig. 4.3). A magnificent high voltage source, the Van de Graaff generator goes mostly unmentioned in the history of particle physics. This was a generator that rivaled and exceeded the CW generator even in 1932. Built by an American physicist Robert Jamison Van de Graaf from Princeton University, it achieved 1.5 million volts in 1931. While there are many types now, in its basic form, the Van de Graff generator employs charges, created on an endless belt by rubbing on a lower roller surface and then carried on and conveyed to a large dome connected to the upper roller. This way, the upper dome continues to build charge and potential with respect to the lower roller. Pointed conical combs can additionally kick off charges into the roller and enhance the effect. In 1933 another generator, built at the Massachusetts Institute of Technology, achieved over 5 MV. In 1937, a Van de Graaff generator, with capacity to be charged also to 5 MV but with a much larger dome (3 m) to store large amount of charge was installed at the newly opened Palais de la Decouverte in Paris, for the purpose of producing radioactive isotopes. Set on 14 m high poles, this was a spectacular machine that could throw a spark to a distance of several meters. World War II intervened and this machine was never used for any serious experiment. While this type of generator and CW generator's kin the Marx generator were also available around the time as the CW generator and are and were used for various nuclear and nonnuclear experiments, these did not participate in the momentous discoveries of physics, such as the cracking of the nucleus and the first transmutation of an element.

The cracking of the nucleus opened up the realm of nuclear physics and laid the foundations for nuclear power and nuclear weapons. We are reaping the good and bad consequences, as societies and nations benefit from the irresistible technologies arising out of these earth-shaking discoveries, but also fear the destructive power unleashed by these discoveries. Because of the fascination with nuclear research, high voltage research too became fashionable among researchers. The public's imagination was being fired up with possibilities in transmutation, splitting of the nucleus, energy sources, and weapons. Across the Atlantic in the USA, a "revolution" was taking place concurrently.

Chapter 5
The Spiral Path to Nirvana

As every amateur electrician who has caused a distressing short circuit at home knows, high voltages tend to spark over. In as much as the Cockroft–Walton generator was admired as the first in a chain of accelerator developments, it was well recognized that such a generator would be limited in its maximum accelerating voltage because of the high voltage capacity of its switches and capacitors. Such accelerators were also limited in their maximum acceleration capability because the particles experience the high voltage at once, rather than in stages. In any case, even generators such as Van de Graff generators were ultimately limited by electrical breakdowns at high voltages. On a good dry day, one could get only a few million volts. A very clever development was taking place in the USA at this time, growing out of a European invention.

In the 1920s, a Norwegian engineer named Rolf Wideroe, working in Karlsruhe, Germany, developed the idea for an accelerator in which the particles, instead of being accelerated by one large electric potential at one end, would be given a small kick of energy every time they passed through a metal tube. His Doctoral thesis was rejected by the Karlsruhe Polytechnic, but was published in Archiv für Elektrotechnik. Many years later, Ernest Lawrence would read this thesis and develop his ideas for a "Cyclotron." Lawrence himself said this often and explained why Wideroe was so popular in the USA. During the Nazi years, Rolf Wideroe and his brother Viggo, an aviation pioneer, were hounded by the Gestapo, but still Wideroe's spirit was never quelled, and both went on to have many more accomplishments. As stated in later chapters, Rolf Wideroe was also the originator of the modern accelerator-storage ring-collider concepts. Once rejected for a Doctorate, later he was awarded an honorary Doctorate by Rheinisch-Westfläische Technische Hochschule (RWTH) in Aachen. In 1964, he even received an honorary medical doctorate from Zurich University along with many other distinctions [*The-infancy-of-particle-accelerators-life-and-work-of-rolf-wideroe*, Pedro Waloschek, Friedr. Vieweg Verlagsgesellschaft, Braunschweig and Wiesbaden, Germany (1994)].

The idea for a resonant acceleration of particles with an alternating field came to Wideroe from the normal transformer. The electrons in a secondary circuit of a step-up transformer can be accelerated to a high voltage while the electrons (currents) are flowing in a copper wire. One just had to get rid of the copper wire and replace it with a circular vacuum tube in which electrons or other charged particles would orbit. The primary circuit would then accelerate the particle at every turn. One, of course, needed a magnetic field to guide the particles into an orbit. The idea was rejected by a physics professor, Goede, in Karlsruhe, who overestimated particle loss through gas–atom collisions. When he did build a machine in Aachen, the machine was too primitive (in his own words) for the time. The electron orbits were unstable because the "Betatron" physics of orbiting particles (see later part of the chapter) had not yet been developed, and there were also problems related to electrons hitting the walls of the tube and causing emission of secondary electrons. So Wideroe gave up this line of research, as this too was a dead end for his Doctoral thesis. Later, accelerators similar to this "ray transformer" would come to be the norm.

Wideroe then followed a different path, pursuing an idea by Gustav Ising, of a traveling wave-type accelerator. He realized that the concept, though not workable directly, was feasible with modifications. Wideroe's device is as follows (Fig. 5.1): Metal tubes were connected to an oscillating voltage, high (radio)-frequency source such that as a particle (shown positive here) enters tube A (Phase 1), the polarity of the tube would be negative to accelerate the particle. As it exited tube A, the polarity of the tube would be synchronized to switch to positive so that the particle would be repelled forward toward tube B, which would have an attractive voltage at that time point. Then, as the particle exited tube (B) (Phase 2), the voltage of tube B would switch polarity, become repelling, and the tube in front (C) would be accelerating (Phase 3). The particle would be accelerated in the gaps and would just (mostly) drift when inside the tube. Therefore, though each tube is maintained at the same voltage, the traveling particle would always be accelerated by the frequency of the alternating potential resonating with the arrival or departure of the particle from the tube.

The energy gain per gap is given by $\Delta E = qE_0TL\cos(\varphi)$, where q is the charge of the particle, E_0 is the peak electric field in the gap of length L, and ϕ is the phase of the electric field wave when the electron arrives at the center of the gap so that the accelerating electric field seen by the particle is equal to $E_0\cos(\phi)$. T is called the transit time factor, which depends on the geometry of the device, particle characteristics, and the RF wavelength. It also accounts for (depends on) the distribution (variation) of the electric field along the gap, the change in phase of

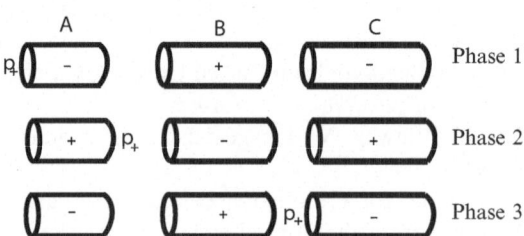

Fig. 5.1 Principle of the accelerator that Wideroe proposed

the electric field as the particle traverses the gap, and also any leakage of electric field into the tube. The transit time factor contains a factor that implies that the faster the particle, the shorter has to be the wavelength (higher the frequency). Since the frequency of the supply is the same for all the tubes, the particles need to spend the same amount of time and stay in phase with the electric field, even as they increase their speeds. Therefore, the (downstream) drift tubes and gaps would have to get longer as the particle gets faster.

The particles would come in bunches (buckets) and each bunch would be spaced two tubes apart so that all bunches see accelerating voltage in front. One could add as many tubes as needed to get the required particle energy. In a historic demonstration in 1927, Wideroe accelerated potassium and sodium ions to 50 keV energy, by using only a 25 kV source, with an 88-cm-long two-tube model. The device apparently cost only about 500 Marks (120 US Dollars in 1927 and 1,500 US Dollars in 2009). There is a small but noteworthy aspect to Wideroe's accelerators; the entrance tube and the exit tube were mere drift tubes at earth potential, which allows all measuring instruments to be at or near earth potential, which was not the case previously. With this device, Rolf Wideroe finally got his Doctor-Ingenieur in November 1927 (Fig. 5.2).

In principle, there is no limit to the energy to which particles can be accelerated using this device. Of course, such a device would get longer (faster than square of the energy) as the energy gets higher, and the ultimate energy would be limited by availability of space. David Sloan in UC Berkeley used this device to accelerate mercury ions to a million electronvolts using 30 electrodes, as early as 1931, but sadly for him, he missed the opportunity to obtain the first nuclear transmutation that Cockroft and Walton achieved. The principle of this linear acceleration is still very popular. A 4-km-long linear accelerator (LINAC) at the SLAC facility at Stanford University in the USA is based on a similar concept (see Chap. 8). However, there was another problem with this type of device, besides the need for a long chain of accelerator tubes. The frequency of the alternating field had to be

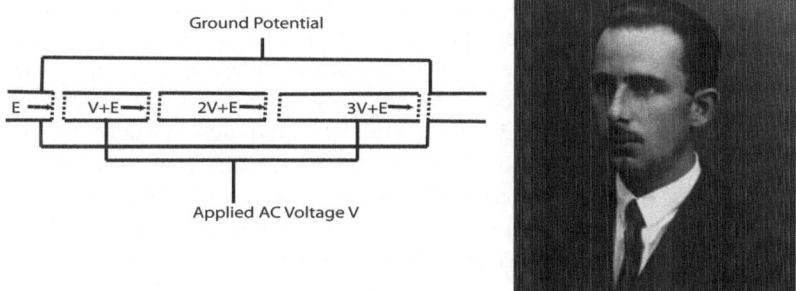

Fig. 5.2 (*Left*) Schematic of Rolf Wideroe's idea for a linear accelerator. (*Right*) Rolf Wideroe during 1930s [The Infancy of Particle Accelerators, Life and Work of Rolf Wideröe, Compiled and Edited by Pedro Waloschek, Friedr. Vieweg Verlagsgesellschaft, Braunschweig (1994), also DESY Laboratory Report DESY-Red-Report 94-039 (March 1994)]

high, in order to accelerate fast particles which would have short transit times. In the late 1920s and early 1930s, high-frequency sources with hundreds and thousands of MHz frequency were not available. This limited not only the energies, but also the type of particles. For instance, light particles such as electrons and protons which would have very high speeds at MeV energies could not be accelerated. High-frequency sources are now available at tens of GHz frequency for LINAC applications.

Homo-Cyclotronus

While the Norwegian-born Rolf Wideroe established one type of accelerator that is in use even today, a grandson of a Norwegian would create the other kind of accelerator, the circular accelerator. Accelerator Nirvana was close. The credit for this goes as much to the aspirations of American academia as to the discoverers themselves. In the words of the American Physical Society, the University of California at Berkeley was being established as "One of the two foci of academic ellipse representing American Physics" (the other being the American east coast institutions). Ernest O. Lawrence, a new breed of physicist who completed all his studies in the USA, was making news with his own important, but somewhat mundane physics experiments on ionization potentials. Amidst a bidding war for recruiting him, Yale University bypassed procedures and made him an Assistant Professor in 1927, at the age of 26 years, the youngest ever. But the University of California won out in the end, with its pots of research money and the offer of an Associate Professorship. There, Lawrence established an iconic particle physics laboratory and presided over many important discoveries. With his invention of the Cyclotron, Lawrence would go on to make the USA a contender for world leadership in the field of experimental physics. Like many things in science, a nexus of ideas and friendly but fierce competition led to these developments (Fig. 5.3).

In the same year as the world was agog with real and imagined potentials of the discoveries by Rutherford and Cockroft, Ernest Lawrence had conceived of a very impressive idea that defines the approach to present-day accelerators. He realized that Wideroe's concept was the way to go, but the linear accelerator of Wideroe would be limited by space and cost. He then came up with what in hindsight is the obvious solution, a circular accelerator in which the particles, instead of traveling in a straight line, would be made to go on a "merry go round," as he described it. This could be done by bending the path of the particle with a magnet.

In an elementary sense, in a circular accelerator, one would need only two or even one accelerating tube whose polarity would be synchronized to switch between positive and negative, depending on whether the particle is approaching or leaving (see Fig. 5.4). When the particle, in this instance a negatively charged electron, is at the top and approaches an accelerating tube on the left, this tube would have positive voltage to accelerate the electron, while the tube on the right would have negative, and the two tubes would switch polarities after the electron

Fig. 5.3 Ernst Lawrence with Enrico Fermi (*Center*) and J. Robert Oppenheimer (*left*) (Courtesy: Lawrence Berkerey Lab Image Library, Image No. XBD9606-2767.TIF)

has left the tube on the left. While this basic concept would be adopted in more recent synchrotrons, the cyclotron was built with a more elegant idea.

A charged particle moving in a magnetic field, experiences a force transverse to the original motion as well as to the direction of the magnetic field. So, when a particle is moving between and in a plane parallel to a magnet's pole faces, the particles would make orbital motion in that plane. For a fixed magnetic field, the radius of the orbit (r) is proportional to the velocity (v) of the (non-relativistic) particle. This means that the particle would spiral out if it is accelerated in the presence of this magnetic field. But it takes the same amount of time to go around the orbit whatever the velocity (radius), as long as the magnetic field is unchanged (time to go around = $2\pi r/v$, and r is proportional to v). If q and m are the charge and mass of the particle, respectively, and B is the magnetic field induction, the particle experiences a force

$$F = qvB$$

in the magnetic field. As the particle is bent into a circular path with radius r, it would experience a centripetal force mv^2/r, which would be equal to F. Therefore,

$$mv^2/r = qvB \quad \text{or} \quad r = mv/(qB).$$

The time taken to go around one orbit is then

$$T = 2\pi r/v = 2\pi m/qB.$$

Fig. 5.4 Principle of
synchronous ring accelerator

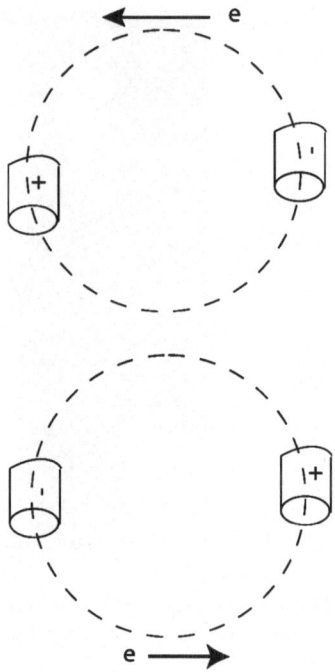

The orbital frequency, the "cyclotron" frequency, is then given by

$$f = 1/T = (qB/2\pi m).$$

As can be seen, this is independent of the radius of the orbit! (For a non-relativistic proton, the orbital frequency is 15.38 MHz/T and about 28 GHz/T for a non-relativistic electron.)

So Lawrence, who read Wideroe's paper of 1928 avidly, came up with the idea of a resonant accelerator, in which, instead of synchronizing the alternating electric field to the linear motion of the particle, the RF field frequency would be resonant with the constant periodic orbital motion which has the frequency given by the expression above. In this scheme, all one has to do is to put the electric field in the accelerating path as in Fig. 5.4. Lawrence and his student Livingstone realized this accelerating pill box in the form of a pair of Dee-shaped chambers placed inside a vacuum chamber, which itself was placed between the pole pieces of a magnet. The two Dee shapes formed the electrodes and the particles accelerated in the gap (Fig. 5.5).

As the particles are injected into the pill box and accelerate, they spiral out to a larger radius – the orbital radius is initially smaller since the velocity is smaller (see equation above), and as the particle accelerates, the orbit radius would increase proportionally. Even as this happens, the resonance condition, in which the orbital frequency is independent of the radius, assures that the particles remain synchronous with the RF field. The condition guarantees that if the electric field is

Fig. 5.5 Principle and scheme of cyclotron, magnetic field B bends particle path into a circular orbit and an alternating electric field applied across two "Dees" accelerates particles

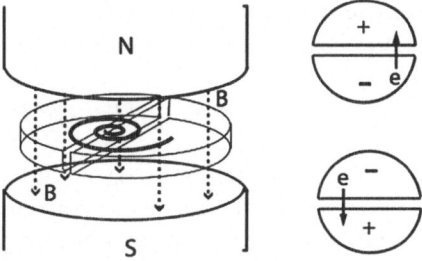

Fig. 5.6 South Indian Folk Dance "Pinnal kolattam" performed by girls in a dance recital (Courtesy: Rajee Narayan and Jayashree Rao, India)

accelerating when the particles enter the gap from one Dee to the other, then as they turn around and return to the first Dee, the polarities of the electrodes are reversed synchronously due to the resonantly alternating radiofrequency source, and the Dee gap remains accelerating (Fig. 5.5). Therefore, the particles are repeatedly and unfailingly accelerated each time they cross the gap (The dance of the particles in a cyclotron is reminiscent of the "Pinnal kolattam" dance in the State of Tamil Nadu, India (Fig. 5.6), and the "May Pole" dance of ancient Britons. In India, the dancing girls would tie one end of their long upper garments or hold a rope tied to the poles to a pole and dance around it in a circle while holding colorful sticks in their hands and hitting neighboring dancer's sticks. They would wind their garment tightly around the pole and slowly unwind as they dance faster and faster, moving to a larger and larger radius as they dance).

The frequency of the *RF* source would be chosen by choosing the strength of the magnetic field and the type of particle. The particles are then extracted (by placing an outlet along the tangent of the orbit) before it hits the walls of the pill box. If the particles are extracted at a radius $R = mv_{final}/(qB)$, where v_{final} is the velocity of the particle at extraction, then the final energy of the particle when extracted is given by

$$E_{final} = mv^2_{final}/2 = (qRB)^2/(2m) = 1.23 \times 10^{20} f^2 R^2 m \, eV,$$

where R is in meters and m is in kilogram. It is worth noting that the final energy is proportional to the square of the radius of extraction (approximately, the radius of the pill box) and the square of the magnetic field (Note also that the final energy is proportional to the mass of the particle, a fact true even today and exploited to achieve hundreds of teravolts of particle energy.). High energies would require high magnetic fields and since the resonance condition demands high frequency at high magnetic fields, one is still limited by the available frequencies of the power sources. The alternative is to allow for large orbital radii, which require increasing the diameter of the pole pieces and this, in turn, would increase the weight and cost of the magnet and power consumed by the magnet, proportional to a power higher than R^2. The orbital frequency is inversely proportional to mass, and therefore, to stay within available source frequencies, low magnetic fields would have to be used to accelerate electrons. Ultimately, the technology and costs limit the RF frequency f and the radius of extraction R. Therefore, electrons would have final energies 2,000 times smaller than that of protons.

Lawrence, like any one in particle physics, was worried about obtaining ultra high vacuum, to avoid particle loss due to collisions with gas atoms. He had encouraging words from Otto Stern, who was an expert on proton beams, from the University of Hamburg. Over dinner during his visit to the laboratory, his enthusiasm over the chance of success in overcoming the vacuum issue stimulated Lawrence to go for it. The final push came from the competition. Ever the rivals, one of the groups on the east coast, Merle Tuve and company, working at the Carnegie Institution in Washington, predicted that they would soon have a 5–10 MV source from a Tesla coil (resonant transformer) with 120 sparks/s, providing radiation equivalent to 2.6 kg of radium.

Though Lawrence's first love was this project, he was obliged to do research on photoelectric effect. Fortunately, he had three very competent graduate students and colleagues – "…a relative importance… that I could never have attained in Yale for years," he wrote to his parents. Hence, Lawrence entrusted the preliminary work to a reluctant graduate student and teaching assistant, Nels Edefson, who built crude Dees out of copper foils glued to glass plates with wax and put in a filament source at the gap. It was the equivalent of the present-day chewing gum and duct tape device. Around the time Edefson left to become a professor of irrigation in the summer of 1930, Lawrence and he claimed that the concept of resonant acceleration had been demonstrated, although this claim was much doubted. However, 6 months later, in the fall of 1930, Lawrence and Milton Stanley Livingstone built a Cyclotron with 4 in. or about 10 cm diameter. The demonstration machine had to overcome many obstacles: obtaining a robust proton source that could provide a mA of current so that at least a micro-amp would survive, obtaining a vacuum of better than 10^{-5} mmHg, a detector cup that collected only high-energy particles at the end of acceleration, and a deflector to corral the accelerated particles into the cup. After many frustrating attempts and many misleading results, Livingston's diligence paid off, and on January 2, 1930, he got a new year gift of a resonant peak at about 1,000 V of gap voltage and 1.24 T (maximum field of the magnet) corresponding to a resonant frequency of about 10 MHz for a charged hydrogen molecule (H_2^+). (When the frequency is such that particles are in phase with the electric field and when many particles are accelerated coherently, the source is "tuned" and one gets a "resonant peak" corresponding to the accelerating voltage, as detected from the source output.) The radius of collection (cup location) of 4.8 cm corresponded to an energy of 80 keV – 80 times the acceleration voltage. The Cyclotron had come into existence! This little cyclotron, which would fit into the palm of a hand, would change the face of accelerator science, radiation therapy, and solid state physics that ushered in the present-day technology boom. In the meantime, David Sloan built the 30-electrode, 1.26 MeV Linac with 10 nA of Mercury ions. Sloan's machine also demonstrated, for the first time, the occurrence of automatic "phase" focusing or longitudinal focusing (see Chap. 7).

During all these, Lawrence was setting a new trend in a different arena. In those days, though consultations were common, experimental physicists worked alone. Lawrence demonstrated how physicists could organize and collaborate among themselves. The "Cyclotron Man" (as Time called him later) became a savvy project leader, promoting healthy competition along with cooperation among physicists. His leadership was so much appreciated that when Northwestern University tried to lure him away in early 1931, Lawrence ended up with a bargain, getting a full Professorship. While all this was going on, Milton Livingston and David Sloan went on to build a 11-in. (26 cm diameter) cyclotron under Lawrence's supervision. They broke the 1 MeV barrier with this compact machine at an extraction radius of just 11.5 cm, with an oscillator source frequency of 20 MHz, which was pretty much the limit in those days. The story goes that during a trip to the east, Lawrence got a telegram from the lab secretary which read "Dr. Livingston has asked me to advise you that he has obtained 1,100,000 V protons. He also suggested that I add, Whooppee!" The success of the Cyclotron owed much to the diligent and careful work of Livingston who tweaked oscillators to get the best performance and shimmed magnets to get the required uniformity of better than 0.1%. Europeans like Cockroft used to say that it was Livingston who did all the work, because Lawrence spent all his time bossing people around. Livingston was a great experimenter, but was not as good at theory. Hans Bethe, the famous theorist, used to explain things to him and in exchange got his only experimental paper.

In this saga of early invention of accelerator designs, two names did not get the credit they deserved. One was Gustav Ising, from whose paper Wideroe got his idea. Of Ising, Lawrence said in his Nobel lecture,

It was only after several years had passed that I became aware of Professor Ising's prime contribution. I should like to take this opportunity to pay tribute to his work for he surely is the father of the developments of the methods of multiple acceleration.

The other might be Jean Thibaud, a French researcher who claimed that he got the resonant accelerator working in 1930, before it was announced at Berkeley. He had, in fact, obtained 1 µA of beam current, 100 times that of Livingston's cyclotron, using differential vacuum pumping of the particle (ion) source to operate it at higher pressure.

Increase to higher energies means increasing the orbit size (because the magnet field strength is limited by saturation of iron). This means bigger magnets and more powerful RF sources. More money was needed and this was, after all, the time of Depression when all budgets were getting slashed. Lawrence had to find other sources of funding for making larger cyclotrons. Two companies, Research Corporation and Chemical Foundations, provided most of the funding for his further work. During this time, there was much wheeling and dealing, many patenting tricks, and several court cases against companies trying to profit from scientific results. The profit-seeking companies knew that high-energy electrons were the need of the hour, and Lawrence bravely talked about grand machines. He proposed a 27-in. (65 cm) cyclotron, using the magnet that was a World War I surplus. Working with Leonard Fuller, a UC Berkeley electrical engineer and vice-president of the Federal Telegraph Company which owned it, Lawrence received the 80-ton magnet, valued at $25,000, as a donation. An entrepreneur-turned physicist, Frederick Cottrell helped steer Lawrence through the deals. Thus, Lawrence got the large sum of $12,000 for modifying, transporting, and installing the 80-ton magnet, for his 27-in. (65 cm) dream cyclotron. The room that he and his colleagues occupied would no longer be enough, as the magnet would need a strong floor. An adjacent 2-story wooden building, which had housed a civil engineering test laboratory (CETL), had a concrete floor. In August 1931, the University President awarded Lawrence full use of the CETL, which was then named "Radiation Laboratory," which became Lawrence Berkeley National Laboratory, today a famous landmark in San Francisco bay area and a center for science. (Lawrence had the unique honor of receiving his Nobel Prize, in view of his beloved Berkley laboratory.) The Cyclotron was operational in September 1932. It produced 3.6 MeV protons, and in 1937, this was upgraded to a 37-in. accelerator with a corresponding increase in final particle energy (Fig. 5.7).

The depression years had set in and UC Berkeley was tightening its belt, slashing a third of its budget. But even in this discouraging environment, Lawrence remained positive and doggedly pursued the drive for bigger and better Cyclotrons. The final thrust of development was that of a 60-in. cyclotron with a 220-ton magnet which was enthusiastically funded by the State of California, Macy and Rockefeller foundations, and others. Lawrence persuaded the community to support his work, prophetically pointing to biomedical applications such as radiation treatment for cancer. Lawrence truly believed in the potential of these particle rays to rival X-rays in their biomedical applications. Since donor organizations gave more money to medicine and biology than to physics, this was also a good move on the part of Lawrence and his associates. The Federal Government provided funds through Franklin D. Roosevelt's "New Deal" funds to create employment, and this, in many ways, represented the visionary society that

Fig. 5.7 80-ton (27 inch) cyclotron, with Ernest Lawrence and Milton Livingston, (Courtesy: Lawrence Berkeley National Laboratory Image Library, Image XBD200904-00162.TIF)

America enjoyed in the depression and post-depression era. Recruitment expanded the physics and engineering staff of the laboratory, and the community's activism was responsible for creating a countrywide interest in science. The 60-in. cyclotron (Figs. 5.8 and 5.9) was completed in 1939, delivering deuterons of 16 MeV energy. It was instrumental in important discoveries such as radioisotopes of helium-3, hydrogen-3 (tritium), radioactive carbon 14, and super-heavy elements 118 and 119. The discovery and isolation of plutonium and uranium were key milestones in the development of nuclear fission devices. These isotopes are the staples of modern science and technology. All these developments made the cyclotrons very attractive. By 1940, there were over 20 cyclotrons in the USA, with similar numbers in the rest of the world. These cyclotrons were even prototypes for many cyclotrons, such as the Variable Energy Cyclotron in India built in as late as 1980.

The cyclotron is a very special invention in many ways. It provided a very robust and yet simple accelerator for energies corresponding to the particle speed of a few percent of light speed (several millions of electron volts of electrons). The radiofrequency oscillators, the vacuum in the pill box, the magnets, and other components could be obtained readily. The whole assembly is relatively compact, providing for easy radiation protection. To this day, these features make the cyclotron most suited for making radioisotopes for hospitals. The original concept of cyclotrons has been refined and made sophisticated, while cost-effective designs have led to commercializing the cyclotron for medical isotope production. Many hospitals have cyclotrons in their basements (Fig. 5.10). The largest cyclotron today

Fig. 5.8 The 60-in. Berkley Cyclotron (left to right) Donald Cooksey, D. Corson, Ernest O. Lawrence, R. Thornton, J. Backus, and W. Salisbury, and (*on top*) L. Alvarez and E. McMillan. (Courtesy: Lawrence Berkeley National Laboratory Image Library, Image XBD9606-02528.TIF)

Fig. 5.9 Accelerated ion beam from the 60-in. cyclotron, allowed to escape to the atmosphere, which ionizes and emits light due to the passage of ions. (Courtesy: Lawrence Berkeley National Laboratory Image Library, Image XBD9606-02749.TIF)

Fig. 5.10 An Advanced Cyclotron Systems 30-MeV variable energy, 1.2-mA negative ion cyclotron for isotope production (Courtesy: Advanced Cyclotron Systems, Richmond, B.C., Canada)

is the 18-m diameter, 520 MeV TRIUMF facility in Vancouver, Canada (see Fig. 5.11), which houses one of the only three "meson factories" that produce beams of the exotic meson particles. Very importantly, the cyclotron displaced all other devices in its preeminence over other concepts, and laid the foundation for all future circular accelerators.

The Synchrocyclotron and Isochronous Cyclotron

Lawrence was very interested in a new class of particles called mesons which had been discovered in the cosmic rays and were suspected to be mediators of the nuclear force. These were quizzical particles that did not fit into the mold of physics then and were somewhat elusive (see Chap. 6). Lawrence desired to create mesons in the laboratory by hitting a light nucleus with a beam of alpha particles. He argued for a cyclotron that, he said, would provide accelerated particles to probe the nucleus so that nuclear forces could be better understood (see previous chapter). The estimated rest mass of these particles was 75 MeV (energy equivalent of mass from $E = mc^2$), and one needed accelerators with somewhat higher rest mass than that in order to create them. Lawrence thought that 150 MeV would be the threshold

Fig. 5.11 Experimental hall of the 520-MeV TRIUMF cyclotron facility, the cyclotron is at the top end of the picture covered by yellow shielding blocks. (Courtesy: TRIUMF, Vancouver, B.C., Canada)

for production of these particles so that even after sharing between the protons or neutrons (nucleons) in the nucleus, there would still be enough energy for producing the mesons.

At these energies, there was need for a new invention. In working with a fixed accelerating frequency, the advantage of cyclotrons is lost as the speed of the particles approaches even a percent of the speed of light. According to Einstein's Special Theory of Relativity, the mass of the particle is related to its speed by the relation

$$\text{Energy } E = mc^2 = \gamma m_0 c^2; \ \gamma = \frac{1}{\sqrt{1 - \frac{v^2}{c^2}}} = \text{relativistic factor,}$$

where v and c are the speed of the particle and light, respectively, and m_0 is the invariant rest mass of the particle. The relativistic increase in effective mass, as the particles approach the light speed, causes the particles to slow down in their orbit relative to the applied frequency and fall out of resonance with the accelerating voltage so that they do not get accelerated any more (The orbital or cyclotron frequency of a relativistic particle is given by $qB/(2\gamma\pi m)$ and therefore reduces as γ increases.). The only way to overcome this is to change (reduce) the frequency of the applied accelerating voltage gradually to be synchronous with particle orbits, as the particle becomes relativistic. The concept of obtaining "phase stability" by adjusting the frequency of the accelerating voltage to match the increasing energy was invented and proved in 1944/1945. The inventors were Vladimir Veksler, at the Dubna Institute for Nuclear Research, and Edwin McMillan, a former student of Lawrence at the University of California in Berkeley. Their invention proved that there was no theoretical limit to the energy to which particles could be accelerated. A cyclotron using synchronous acceleration by frequency modulation (FM) has usually been called a synchrocyclotron or an FM cyclotron. Accurate tuning of the frequency creates higher particle stability, and the particle losses in each orbit are reduced. The advantage gained by synchrocyclotrons is that since the orbit stability is greater, the particle can have many more orbits and the accelerating voltage between the two Dees can be reduced to impart smaller energy steps to get to the same energy. Typical voltages are about 10 kV as opposed to 30 or so kilovolts in a cyclotron. An alternative has since been developed in which, to compensate for the reduction in the orbit frequency, the magnetic field is increased. These "isochronous cyclotrons" have replaced the synchrocyclotrons.

Later, Lawrence raised his estimate of the energy required to create mesons. While Lawrence did his marketing campaign convincing the funding agencies that this energy would be sufficient, Edward Teller and W.G. McMillan encouraged the construction of this type of machine because their estimates found that such alpha particle energies might be enough to produce copious amounts of mesons because at higher energies, the nucleus would be quite transparent and all the energy was more likely to be absorbed by a single nucleon (proton or neutron) rather than being shared among many nucleons. This would then lead to a high probability of creating a meson. In October 1939, Lawrence discussed a giant 184-in. cyclotron with Warren Weaver, Director of Rockefeller Foundation. After many go-arounds, the final proposal was submitted in April 1940. In the process, Lawrence battled many doubters and critics but the machine was finally approved.

Lawrence Radiation Lab's 184-in. synchrocyclotron (Fig. 5.12) produced its first beam on the midnight of November 1, 1946. At a grand celebration of this huge device which doubled the electricity consumption at the new site on Radiation Lab hill, the luminaries and guests from contributing corporations and philanthropic organizations, along with government scientists and administrators, watched as the needle on the energy dial of the 184-in. cyclotron pointed to 195 MeV deuterons. The 4,000-ton machine would be described by journalists as "atom-smasher," as though the laboratory would smash atoms by dropping the machine on them. In any case, the machine vindicated Lawrence, Teller, and McMillan when Gardener and Lattes produced the pions, very important members of the meson family. Under the leadership of Lawrence, the University of California, Berkeley, had, in a matter of a decade, pushed the limits of particle energies hundredfold.

The synchrocyclotron was recruited into the war effort, in a refurbished machine called Calutron. The Calutron, using an electromagnetic separation method, produced part of the enriched U^{235} for the Manhattan atom bomb project. The largest synchrocyclotron 1-GeV (1,000 MeV) machine ever built and still operating is located in Gatchina outside St Petersburg. The whole accelerator weighs 10,000 tons and has a massive magnet of 6-m (240 in.) diameter pole piece. Edwin Mattison McMillan received the 1951 Nobel Prize in Chemistry together

Fig. 5.12 Ernest Lawrence and team at the massive 184-in. cyclotron at the Berkeley Radiation Lab (Now called Lawrence Berkeley Lab). Image No. 96602772, Lawrence Berkeley Laboratory Image Library

with Glenn Theodore Seaborg for the discovery of the element neptunium using a synchrocyclotron.

In 1958, the European high energy laboratory CERN's first accelerator, the SC synchrocyclotron (see Chap. 7), made one of the most important discoveries in physics by demonstrating the decay of pions into electrons and thereby confirmed the theory of weak interaction force, one of the nature's four fundamental forces that causes radioactive nuclei to decay, stars to shine, and made nuclear reactors possible.

While the Radiation Lab kept its nose to the grinding wheel of accelerators, the field of physics was undergoing tectonic shifts. Quantum Mechanics was becoming the "in" field, arousing passions on both sides of the fence while sparking great imagination and driving discoveries. Field theory was becoming prominent. New particles were being discovered everywhere. Proving Rutherford wrong, many discoveries did not need man-made accelerators.

Betatron: The Particle Shepherd and Mother of New Inventions

While the cyclotron was a major breakthrough bringing many laurels to the inventors and resulted in significant applications to physics and technology, the maximum energy was ultimately limited by the limitations on acceleration due to relativistic mass effects. At 1 MeV, the orbital frequency of an electron would reduce by a factor of two. Therefore, at higher energies, it would be hard to keep up with the changes in required accelerator frequency. So, while Lawrence was working on his Cyclotron, Donald Kerst worked on a parallel concept on accelerating the particles inductively. This method, originally but incompletely conceived by Wideroe in 1923, cleverly used the basic induction principle of electromagnetism to generate an electric field by maintaining particles (electrons) in a tight fixed-radius orbit with a steady magnetic field, and applying a varying magnetic field inside the orbit (see Maxwell's equations in Chap. 3). In this situation, the particle always experiences an accelerating electric field all along its orbital path, so no high-frequency source would be needed, but one would need to change the magnetic field as the particle accelerated. There are two important characteristics to note here – one, the orbit has a fixed radius, not a spiral one, and the other, acceleration is due to an inductive electric field produced by a changing magnetic field (Fig. 5.13).

It is well known that when a magnet is moved in and out of a loop of wire, a voltage is induced according to Maxwell's law – also called Faraday (inductive) voltage, crediting Faraday who demonstrated this effect. The faster the magnet moves, the greater is the induced voltage in the wire. The same effect is obtained when the strength of the magnetic field is changed. As the magnetic field is increased, the particles experience this inductive voltage which acts to provide the accelerating force in the tangential direction.

As shown in Fig. 5.5, as a particle accelerates, it would spiral to a larger radius, for a fixed magnetic field. But, in a miracle of nature coming together for our benefit, in the case of the Betatron, the increasing magnetic field exactly compensates for the acceleration, and the particle stays in the same orbit. That is, the change in momentum (mass × velocity) is always proportional to change in (average) the magnetic field. Moreover, since the orbit radius ($r = \gamma mv/eB$) depends on the ratio of momentum to the magnetic field, the orbit radius remains fixed. The magnetic field variation across the magnet pole has to be adjusted to a profile that is appropriate, in order to maintain the particle at the desired radius (see below). To obtain this desired radius, a fixed magnetic field is also applied, providing flexibility for the changing magnetic field. Of course, one cannot

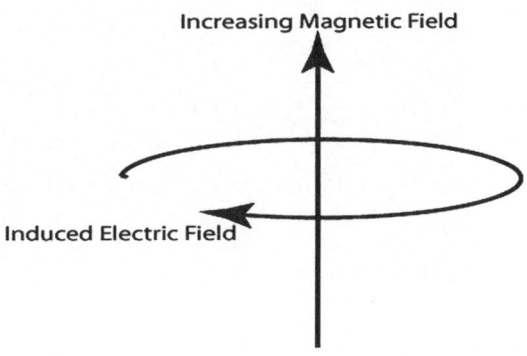

Fig. 5.13 Applied varying magnetic field and induced accelerating voltage

continuously increase the magnetic field and so once the magnetic field is raised to the maximum value, the acceleration has to stop and the particle ejected.

While Wideroe had unsuccessfully tried the concept of Betatrons nearly two decades before, Kerst was the first to build a 2.2 MeV-machine in 1940, after which he and his team went on to build successively larger machines – a 20-MeV machine in 1942, a 100-MeV Betatron in General Electric, leading to the massive 300-MeV machine completed in 1950 (Fig. 5.14). Kerst's singular success was due to the fact that he carefully studied the theoretical aspects of particle dynamics and paid careful attention to all the subsystems that would be part of the accelerator, such as the magnet, the power supply, the vacuum pipe, and the vacuum system. Therefore, he not only advanced the state of art of accelerators and increased attainable particle energy, but also laid the foundation for a systems and quality conscious approach to the construction of large accelerators, which is now part of the culture of accelerator builders.

One important point needs to be mentioned. The cyclotrons could not accelerate electrons to high energies because these would need high-frequency *RF* sources. But, betatrons are excellent electron accelerators, because they do not need *RF* sources. In addition, betatrons are bad ion accelerators. The ratio of final energy that can be achieved for an ion to that which can be achieved for an electron is given by

$$\frac{E_{\text{ion}}}{E_{\text{electron}}} = \frac{v_{\text{i}}}{v_{\text{e}}} \sim \frac{v_{\text{i}}}{c},$$

Fig. 5.14 The 300-Mev Betatron magnet of University of Illinois, Chicago, being placed in position (June 1949) Credit: Carl Pittman, University of Illinois, Urbana Champagne, IL, USA

where v_i and v_e are ion and electron velocities, respectively. This is because for a given energy, an electron is much faster than an ion (by a factor = square root of the mass ratio). The magnetic field has to be ramped over a specific time to get a certain electric field since the electric field is proportional to the rate of increase in magnetic field strength. Therefore, the accelerating voltage times the duration of ramp is fixed. So the particle that can complete more turns in the duration of magnetic field wins the energy race. This can be seen in a back of the envelope analysis.

$$V \sim dB/dt \sim B/T_R,$$

where T_R is the time over which the magnetic field is ramped. The final particle energy

$$E_{final} = V \times N \sim BN/T_R,$$

where N is the number of orbits the particle makes in the time $T_R = T_R/T$, where T is the time for one orbit (period) $= 2\pi r_{orbit}/v$, where r_{orbit} is the orbit radius and v is the particle velocity.

Therefore, $E_{final} = B/T = Bv/2\pi r_{orbit}$.

Following the above expression, if a proton can be accelerated to 1 MeV (proton velocity is $\sim 10^7$ m/s) in a betatron, the same betatron (same magnetic field and orbit radius) can accelerate an electron to about 30 MeV.

Betatron Physics

One of the basic rules that Kerst obtained for the Betatron was that there is a condition that needs to be maintained for achieving fixed orbit radius, while the magnetic field is increased. As noted above, the increase in magnetic field does two things: first changing magnetic field induces the accelerating electric field, which accelerates the particle, and second, the increased magnetic field keeps the more energetic particle in the same orbit radius r_{orbit}. One obtains the "betatron condition" by taking care of both requirements.

The electric field E (tangential to the particle orbit) is induced by the magnetic field change and the associated change in magnetic flux $\Phi = \pi r_{orbit}^2 B_{av}$ (B_{av} is the magnetic field averaged over the pole width) and is given by

$$E = -(d\Phi/dt)(1/(2\pi r_{orbit})) = -(r_{orbit}/2)(dB_{av}/dt). \tag{5.1}$$

The force F experienced by the particle due to this electric field is qE, where q is the electric charge of the particle.

Or

$$F = -(qr_{\text{orbit}}/2)(\mathrm{d}B_{\text{av}}/\mathrm{d}t) = \mathrm{d}p/\mathrm{d}t \ (\text{by Newton's law}). \tag{5.2}$$

Integrating

$$p = -(qB_{\text{av}}r_{\text{orbit}}/2), \tag{5.3}$$

where p = particle momentum = $-\gamma m v_\theta$, where v_θ is the particle velocity in the direction tangential to the orbit (azimuthal velocity). (Sign of the velocity is chosen such that it is consistent with the direction of an accelerating electric field.). But,

$$r_{\text{orbit}} = (\gamma m v_\theta)/(qB_{\text{orbit}}) = p/(qB_{\text{orbit}}), \tag{5.4}$$

where B_{orbit} is the magnetic field at the orbit radius.
Or

$$p = -qB_{\text{orbit}}r_{\text{orbit}}. \tag{5.5}$$

Comparing (5.3) and (5.5) one gets the betatron condition,

$$B_{\text{av}} = B_{\text{orbit}}/2, \tag{5.6}$$

that is, the average magnetic field should be twice the field at the orbit radius.

This elegant but stringent condition requires a magnetic field that varies along the radius. But how it is varied determines the stability of the orbit. As the energy is increased and orbit radius remains the same, two well-known phenomena come into play. The first is the so-called betatron oscillations and the second one is the problematic-helpful-crucial synchrotron radiation. Both these effects are more pronounced for electrons than for heavier protons and ions.

In order to maintain the particles in a stable orbit vertically and horizontally so that particles do not stray away, the magnetic field is shaped carefully in Betatron so that any straying creates a force that pushes the particle back into the orbit. This method is called weak focusing. The vertical focusing of the circulating particles is achieved by varying the applied magnetic field along the radius in a particular way, decreasing from inside to outside. At any given moment, the average vertical magnetic field sensed during one particle revolution is larger for smaller radii of curvature than for larger ones. This also means that the particles are allowed to stray a little bit (both vertically and horizontally) before they experience the force (like a straying child called back by his mother). This movement, in vertical and horizontal direction, is correlated and the particles make a small spiraling motion around the orbit. These are called betatron oscillations and the characteristics of these oscillations are even now studied by accelerator physicists. Weak focusing is a concept that is fundamental to particle dynamics and was continued in use for this and the next generation of accelerators.

The fact that the particles in motion in a magnetic field would experience an oscillatory motion can be understood by all who have thrown a ball down a wide concave gutter. The ball would ride up and down the sides of the gutter as it goes along the gutter at a slight angle. A similar visual can be seen (Fig. 5.15) with a luger sledding down the luge track while banking on the sides of the track where the sled would oscillate back and forth perpendicular to the track. The radial magnetic field is zero at the center and increases away from the center. The force from the radial magnetic field pushes the vertically straying particle toward the center. This is similar to the rising sides of the gutter and gravity pushing the ball back to the middle of the gutter. If the ball always has a nonzero sideways speed, it is going to resist this inward push and rise up till the inward push by gravity slows it down and then the ball would fall back now gathering speed toward the opposite side. Then it would fall back from that side, oscillating back and forth. The particle does the same in the vertical direction.

The oscillatory motion can be seen analytically in the following manner: We saw that the force experienced by charged particles in a magnetic field (force that keeps

Fig. 5.15 An athlete riding on a sled down the luge track. His ride along the sides of the walls would be more or less depending on how sloping it is

the particle bent in an orbit) $F = qvB$, where so far we have assumed that the magnetic field is only in the vertical (z) direction between pole pieces (B_z) and the particle velocity is only in the rotation tangential (θ) direction. In reality, the requirement of a radially varying vertical field means (from Maxwell's equation) that a radial component of the magnetic field is present. The particles too have small velocities in the vertical (z) and radial (r) directions. Therefore, there are additional forces in the radial (moving the particle radially inward or outward) and in the vertical directions. The force in the vertical direction

$$F_z = qv_\theta B_r \sim qv_\theta \frac{\partial B_r}{\partial z} z, \tag{5.7}$$

where $\partial B_r / \partial z$ is the gradient of the magnetic field in the radial direction (The radial magnetic field is zero at the center: $B_r = 0$ at $z = 0$, and increases approximately linearly with vertical distance; the sign of the rotational velocity is negative). Since we want to keep the particles confined to a beam, the radial magnetic field has to be in a direction such that it pushes the particle toward the center when the particle displaces in the vertical direction). But the fact that the magnetic field is produced in vacuum means (from Maxwell's equation $\nabla \times B = 0$) that

$$\frac{\partial B_r}{\partial z} = \frac{\partial B_z}{\partial r}.$$

This equation states that if there is a gradient in horizontal (radial) direction, there is also a radial magnetic field (with strength increasing with z).

$$\begin{aligned} F_z &\sim qv_\theta \frac{\partial B_z}{\partial r} z \\ F_z &\sim m \frac{\partial^2 z}{\partial t^2} = \frac{mv_\theta}{qB_z} \frac{q^2 B_z^2}{m} \frac{1}{B_z} \frac{\partial B_z}{\partial r} z. \end{aligned} \tag{5.8}$$

Defining, $\omega_0 = qB_z/m =$ cyclotron (orbit rotation) (angular) frequency, (5.8) can be written [using (5.4)] as

$$\frac{\partial^2 z}{\partial t^2} = \frac{r_{\text{orbit}}}{B_z} \omega_0^2 \frac{\partial B_z}{\partial r} z = -\omega_z^2 z; \quad \omega_z^2 = -\omega_0^2 \frac{r_{\text{orbit}}}{B_z} \frac{\partial B_z}{\partial r}. \tag{5.9}$$

As one can recognize, the equation

$$\frac{\partial^2 z}{\partial t^2} = -\omega_z^2 z \tag{5.10}$$

describes a simple harmonic motion with a frequency of $\omega_z/2\pi$, if ω_z is real and positive, that is, if ω_z^2 is positive, then solutions are of the form $z = z_0\sin(\omega_z t)$.

For ω_z^2 to be positive, $\partial B_z/\partial r$ has to be negative (the main vertical magnetic field has to decrease radially outward). This ensures that the small (value zero at $z = 0$) radial magnetic field is in such a direction (the radial magnetic field points toward the center – has a negative value) as to push the particle to the middle if it strays out vertically. Now, if $\partial B_z/\partial r$ is positive and ω_z^2 is negative, then (5.10) describes a particle displacement given by the form $z = z_0 \exp(\omega_z t)$, an exponentially growing disturbance. Since we would desire a stably oscillating particle motion rather than an undamped large displacement that can lead to particle losses, the machines are arranged to have this negative radial field gradient, and hence a stable oscillatory motion is obtained.

The magnetic poles are shaped (tapered to provide increasing gap with increasing radius, thereby providing a magnetic field which decreases with increasing radius) in order to obtain a stable vertical "betatron" motion (Fig. 5.16). The curved field lines actually reduce radial confinement of particles a little bit, but promote orbital stability in the vertical direction.

The equation can also be written (by dividing the above equation on both sides by r_{orbit}^2) as

$$\partial^2 z/\partial s^2 = (1/(B_z r_{\text{orbit}})(\partial B_z/\partial r)z, \qquad (5.11)$$

where s is the distance traveled along the orbit $= (v_\theta t) = (\omega_0 r_{\text{orbit}} t)$. This describes the oscillatory trace of the particle as it orbits.

A similar analysis for the horizontal (radial) direction also gives an oscillatory motion with a somewhat different equation

$$\partial^2 r/\partial s^2 = -(1/(B_z r_{\text{orbit}})(\partial B_z/\partial r) + (1/r_{\text{orbit}}^2))r. \qquad (5.12)$$

Therefore, a particle traveling around or through a magnet will perform oscillatory motions in the directions perpendicular to its velocity. These are called betatron oscillations, after the fact that these were first established by Donald Kerst for the Betatron. However, the radial betatron motions are performed in a field

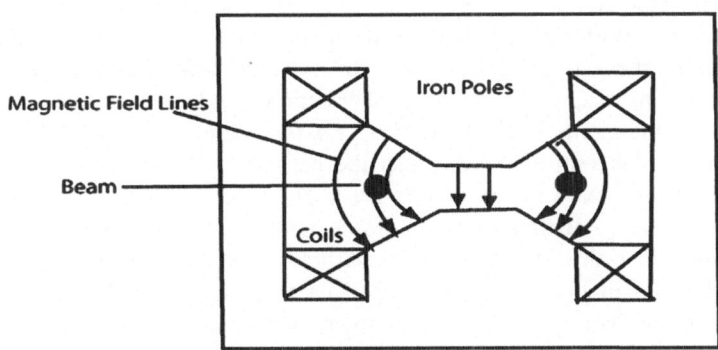

Fig. 5.16 Cross-section of a typical betatron magnet with curving field lines at the beam location

gradient that is opposite of the case of the vertical oscillations. In this case, there is less stability for a negative radial gradient, because the particles, moving out along the radius, move to a lower field region and their orbit radius would increase further moving them out even further (Note that the first term in (5.12) is destabilizing for negative radial gradient.). Some margin of stability in the radial direction is, however, obtained because of the curvature of the path – because of the second term which corresponds to the restoring centripetal force which pulls the particles back.

The specified variation in the field that is required to obtain stable betatron oscillations and established by the shaping of the poles may be represented as

$$B_z = B_{z0}(r_{\text{orbit}}/r)^n, \tag{5.13}$$

where n is the "field index" which determines the strength of the gradient.

$$n = -(r_{\text{orbit}}/B_{z0}) \; (\mathrm{d}B_z/\mathrm{d}r)_{r=r_{\text{orbit}}} \text{ (that is, the gradient is evaluated at}$$
$$r = r_{\text{orbit}}). \tag{5.14}$$

The equations can then be written as

$$\partial^2 z/\partial s^2 = -(n/r_{\text{orbit}}{}^2)z = -\omega_z^2 z \tag{5.15}$$

and

$$\partial^2 r/\partial s^2 = -((1-n)/r_{\text{orbit}}^2))r = -\omega_r^2 r \tag{5.16}$$

For stable oscillation in the z direction (ω_z is real, ω_z^2 positive), n has to be positive [see (5.15)]. However, the radially decreasing field permits larger radial oscillations. For ensuring stable and limited oscillation in the radial direction (ω_r is real, ω_r^2 positive), n has to be less than 1 [see (5.16)], which gives a stringent betatron stability condition, $0 < n < 1$ for "weak focusing" accelerators. (The term weak refers to the fact that the oscillation wavelength is large compared to the orbit circumference). With the radial magnetic field, which pushes the particle to the center keeping them on track, increasing up and down from the center, the particle is similar to being in a bowl in the vertical direction. The larger is n, the larger is this effect. But in the radial (horizontal) direction, the decrease in vertical field is like slope which, if uncontrolled, can push the particle radially out, except that the circular orbit path keeps the particle in leash. However, the gradient should not be so large as to overcome this restraining force, and hence n has to be smaller than 1.

Therefore, in order to maintain the particles in a stable orbit vertically and horizontally, the magnetic field is shaped carefully in a betatron to satisfy the above condition. The quantity "n" is known as the magnetic field index and for considerable period of time, the tight prescription on the field index with shaped poles and specified spatial variation in magnetic field remained the essential feature of particle accelerators for a generation of accelerators including the early synchrotrons. The field index was one of the sacred parameters in the design of accelerators, and designers shaped the magnetic field variation meticulously in obtaining the right field index.

The betatron concept of a circular machine provided the basis for the development of a detailed, theoretical description of particle motion and its stability, in the environment of nonideal conditions, such as magnetic field errors and residual gas pressure, collective effect of the beam (effect of the collective charge of the other beam particles on a given particle), and nonideal accelerating electric field with small variations in frequency and magnitude. As seen in later chapters, even in modern accelerators, betatron motions are design parameters prescribed by accelerator designers all over the machine circumference, and experimental accelerator physicists would measure and fuss about the betatron parameters to determine particle beam behavior.

Synchrotron Radiation

It has been well known that accelerating charged particles emit radiation with the power and energy radiated being high for electrons. Since electrons move in a circular orbit (changing direction continuously, which represents acceleration in a new direction) in cyclotrons and betatrons, they emit synchrotron radiation continuously. Higher the energy of the particle, higher is the energy radiated and smaller is the wavelength (wavelength in nanometers $\sim 1.864/$(magnetic field in Tesla $\times$ energy in GeV^2). Strong synchrotron radiation was first observed on the 70-MeV GE Synchrotron by Elder, Gurewitsch, Langmuir, and Pollack. Langmuir recognized it to be the electron-synchrotron radiation, though the team first called it Ivanenko and Pomeranchuk radiation after the originators of the idea. The emission of synchrotron radiation actually represents an energy loss to the electron, but this "dissipation" helps to damp out the betatron oscillations so that spirals around the main orbit become tighter as the energy of the electron increases, which is very helpful in creating a focused beam.

Electrons acquire near light speed at about 0.5 MeV and therefore at the multi-MeV energies of cyclotrons and betatrons, they are relativistic. When the relativistic electrons go around an orbit, the synchrotron light is emitted mostly in the forward direction. This creates a bright and tight cone of light and the energy is also more or less monochromatic (of single wavelength). This ability to obtain the brightest beams of light in X-ray wavelengths made Betatrons extremely attractive for physics studies. The fact that the electrons stayed in a single orbit meant that the synchrotron radiation could be obtained from a port tangential to the fixed orbit radius. Such high-energy, high-intensity X-ray synchrotron sources have been responsible for many major discoveries in material science, physics, and biology. Accelerator-generated X-rays are now used to study hard and soft materials such as solid-state devices, thin films, and biological membranes, and to develop new materials like the strongest and hardest carbon nano-tubes. In a bread-and-butter activity, many high-tech electronic chips are made by etching using synchrotron radiation. A 25-million-volt betatron was installed at the University of Illinois' College of Medicine in Chicago as the first accelerator dedicated to cancer treatment. The first patient treated with it (TIME, Sept. 5, 1949) was Fordyce Hotchkiss, aged 72 years, a retired Railway Express employee, who had an egg-sized cancer of the larynx. After a few months of treatment, his cancer was pronounced "healed." Synchrotron radiation from modern-day Synchrotrons (see next chapter) are largely responsible for many innovations that have made the present information and medical technologies possible. Accelerator-based synchrotron radiation sources from ASTRID in Denmark to SESAME in Jordan are contributing to innovations in science, technology, and medicine.

While the 100-MeV Betatrons were used to study the structure of the nucleus consisting of protons and neutrons, the much larger Betatron(s) generating 300-MeV electrons and X-rays were used to study the structure of the protons and neutrons themselves. The Betatron was the first accelerator to provide gamma rays for studying interactions between energetic photons and nuclei; for example, photo disintegration of the deuteron (nucleus of the hydrogen isotope), and the discovery of the Giant Dipole Resonances of neutrons oscillating in correlation with protons in a nucleus (predicted by Edward Teller and Maurice Goldhaber) in an excited nucleus (nucleus that is in a temporarily higher energy state).

Chapter 6
Then It Rained Particles

The negative and positive spiritual forces (kuei-shen) are the spontaneous activity of the two material forces (yin and yang)... Chang Tsai, 11th century Chinese Philosopher
If you are not confused, you are not paying attention... Tom Peters, Author, Management books

In the 1930s and 1940s, with people like Rutherford and Lawrence minding the experimental shop and great minds pondering the issue of fundamental particles, there was a relatively happy atmosphere of exploration. Clearly, new physics tools, equations, and methods were available and the exploration of breakthroughs in physics was in the wind. It was inevitable that particle physics would make big strides and make new discoveries. But the pace of particle discoveries became a bit breathless and confusing, and theorists and experimenters had a tough time understanding and tagging these particles. But out of this confusion came one of the most abiding foundations of future physics. Such is the nature of scientific quests.

Seeing Signs: The Negatron and Positrons

Paul Dirac was a genius in his own right. As C.P. Snow said, "(Paul Dirac) was brought up by Swiss parents to be fluent in both French and English and was singularly reticent in both, but eloquent in his manipulation of mathematical symbols." In 1927, this man who would occupy the Lucacian Chair of Mathematics, originally held by Isaac Newton at Cambridge, let his mind speak by creating the field of Quantum Field Theory where the relativistic treatment of continuous (classical) electromagnetic field is changed into a description of how photons behave in a quantized electromagnetic field. Isaac Newton's classical mechanics assumed a Universe in which all properties varied smoothly, and time intervals and spatial dimensions were independent of who was measuring them. But, starting from Planck's work, de Broglie, Heisenberg, and Born had developed the field of

R. Jayakumar, *Particle Accelerators, Colliders, and the Story of High Energy Physics,*
DOI 10.1007/978-3-642-22064-7_6, © Springer-Verlag Berlin Heidelberg 2012

Quantum Mechanics, in which particle momenta, position, and energies and associated electromagnetic fields only vary in discrete steps. Einstein had previously developed the Theory of Relativity in which time intervals (simultaneity of events) and spatial dimensions and particle momenta and energies depended, in general, on the observer's frame of reference. However, the two fields of quantum mechanics and relativistic mechanics remained tangential to each other, until Dirac used the two fields in a stunning way.

According to Einstein, the "classical" (meaning non-quantum mechanical) total energy E of a particle with a rest mass of m_0 and velocity v is given by

$$E^2 = p_r^2 c^2 + (m_0 c^2)^2 \qquad (6.1)$$

(In Newtonian Mechanics, the relationship is very different viz. $E = p^2/2m_0$, $p = m_0 v$.),

where p_r is the relativistic momentum $= \gamma m_0 v$, p is the non-relativistic momentum $= m_0 v$, and c is the speed of light.

On the contrary, the prevalent quantum mechanical description of particle existence and travel (propagation as matter wave) was described by (4.1) as

$$i\hbar \frac{\partial \psi}{\partial t} = E\psi = -\frac{\hbar^2}{2m} \nabla^2 \psi + V(x, y, z)\psi.$$

ψ being the wave function of the particle with mass m, $\hbar = h/2\pi$, h being the Planck's constant, and V is the field potential (energy) in which the particle is moving. Schroedinger had written this equation treating matter as waves and did not represent the physical parameters of the matter in any particular sense. Werner Heisenberg, who had developed the Matrix representation for the same equation, felt he had a better grasp on the matter–wave duality.

John (born Janos in Budapest, Hungary) von Neumann was a child prodigy, able to share jokes in Greek at the age of six and able to memorize phone books. After his talents were carefully nurtured and he had done considerable mathematical work, he became world renowned. He was invited to Princeton to work on quantum mechanics and within only a couple of years, he developed his own version of quantum mechanics based on what he called "operator" theory. This, as if magic, immediately brought together Schroedinger's wave mechanics and Heisenberg's matrix mechanics. In the new and widely accepted quantum mechanical description, the energy is also an operator given by $i\hbar \frac{\partial}{\partial t}$ where i is the square root of (-1), and $\frac{\partial}{\partial t}$ is the partial derivative – an operator that takes a derivative of a function with respect to time (only). Similarly, $i\hbar \frac{\partial}{\partial x}$ is the operator corresponding to the x component of the momentum. These operators are used in conjunction with the wave function and various principles such as Heisenberg's uncertainly principle were given a proper mathematical description (In this operator theory, the operators corresponding to position and momentum do not commute – meaning that the result depends on the sequence of operation, see Chapter 10). With this, one more brick had been laid to the foundation of quantum mechanics.

In this description, the meaning of the operator then is that the result of the operation $i\hbar\frac{\partial}{\partial t}$ on the wave function ψ gives the energy of the wave function multiplied by the wave function. That is,

$$i\hbar\frac{\partial}{\partial t}\psi = E\psi \qquad (6.2)$$

and similarly for the momentum operator. Then what becomes evident is that the Schroedinger equation is the statement that energy and momentum are related by the expression

Total energy $E = p^2/2m$ + potential energy

$$i\hbar\frac{\partial}{\partial t}\psi = E\psi = \frac{p^2}{2m}\psi + V\psi = -\frac{\hbar^2}{2m}\frac{\partial^2}{\partial x^2}\psi + V\psi \qquad (6.3)$$

One-dimensional equation. If we follow the same procedure, the relativistic equation (6.1) can be written as

$$\left(E^2 - p_r^2 c^2 + \left(m_0 c^2\right)^2\right)\psi = 0. \qquad (6.4)$$

However, the non-relativistic relation $E = p^2/(2m)$ has only the first power of energy E (linear in E), while (6.4) has the energy as squared, and therefore if (6.4) were converted into an operator type of equation using (6.2), then the time derivative would get squared, that is,

$$-\hbar^2\left(\frac{\partial}{\partial t}\right)^2\psi = E^2\psi.$$

This would then lead to an equation similar to (6.3) (called Klein–Gordon equation). But such an equation would no longer correspond to a linear operation in time $\frac{\partial}{\partial t}$ that is obtained in (6.3) and would lead to a second derivative $-\hbar^2\frac{\partial^2}{\partial t^2}\psi$, while the momentum operator, still entering as square, would remain the same $\left(-\hbar^2\frac{\partial^2}{\partial x^2}\psi\right)$, as in (6.3). This would then look more like a classical electromagnetic propagating wave equation. This creates considerable difficulties in particle descriptions, with the meaning of the wave function itself getting affected and affecting the meaning of $|\psi|^2$ as the probability (density) of a specific state of the particle (see Chap. 4).

This is where Dirac's genius and bold insight came – In 1928, he stated that in order to preserve the quantum mechanical relationship between wave functions and probabilities of particle states, (6.1) must be written as

$$E = \pm\sqrt{\left(p_r^2 c^2 + \left(m_0 c^2\right)^2\right)}. \qquad (6.5)$$

This allowed the Schroedinger wave equation (6.3) to be rewritten as

$$(E - \alpha_\mu p_r c - \alpha_v m_0 c^2)\psi = 0, \tag{6.6}$$

where the new α coefficients (called spinor matrices) describe the properties of the state of the electron, but now both the energy and momentum enter as first-order operators. When the quantum mechanical description of an electron is obtained from this equation, it promptly gave an additional intrinsic property of "spin," already discovered in experiments, but until then unaccounted for by theory.

A very interesting and startling result one sees in (6.5) is that the electron can have a negative energy. Dirac stated that "(this) corresponds to electrons with a peculiar motion such that the faster they move, the less total energy they have, and one must put energy into them to bring them to rest." He stated further that since this state is not familiar to us, we must find a meaning and certainly should not blindly legislate against it. An examination of these states in the presence of electromagnetic fields shows that such an electron (with a negative energy) would behave as if it has a positive charge. Thus was born the concept of "positron" (named by Carl Anderson). Dirac suggested that when all possible negative states of an electron are filled, the electron would exist as we know it. But if there is an unoccupied negative energy state, a "hole" so to say, then the particle must be a positron (The term "hole" would be prophetic and would get used in solid-state physics to describe electronic vacancies.) (A statement similar to Sherlock Holmes's *"Eliminate all other factors, and the one which remains must be the truth."*). In 1931, Dirac published a paper postulating the anti-electron particle. Eventually, this would be the beginning of an understanding that where there is matter, there must be antimatter. The Universe has Yin as well as Yang.

Tracking the Positron: Carl Anderson

Robert Millikan was the physicist who had set the standard for accurate measurements in particle physics by measuring the charge and mass of the electron, the man who studied Victor Hess's radiation from the sky carefully and identified possible source and named it cosmic rays, and the man who came to inherit the mantle of the leader of upcoming center of physics at the California Institute of Technology (Caltech), only about 1,000 km from where Lawrence's team was making history. Millikan negotiated a deal to be the President of Caltech (then called Throop Institute) with only few administrative duties so that he could continue to be involved in science. He set the theme for the Institute which is in place even today. The focus was on cutting-edge technology with a close association with scientific discoveries. Application of quantum mechanics to chemistry, research in biotechnology such as chemistry of vitamins, radiation therapy, study of turbulence for application to airplane design, and study of geophysics for tracking earthquakes were examples of his far-reaching vision that primed science and technology alike.

His service, wedged between the two World Wars, inspired Nobel prize winning discoveries in physics and chemistry at the Institute. In recent decades, this culture made it the most suitable partner for NASA's Jet Propulsion Laboratory. While up north Lawrence was a man much like the directors of the present-day weapons laboratories – more managers than scientists, Millikan was more like the present-day high-energy laboratory directors, meeting their managerial obligations while not compromising on their personal scientific pursuits and keeping themselves fully knowledgeable about the science they manage. His faith in a basic science-based future of America was unshaken and today, American, indeed the world's, economic, military, and IT leadership springs from the vision of scientists like Millikan. Despite this, Millikan was a man who was against government-funded research and so this man would not fit in today's scientific society, in which 99% of the basic research funding comes from the government and as a result, any research can be cut off at its knees if it did not conform to the politics of the moment.

Under Millikan's ownership and leadership in cosmic ray physics, the next experimental breakthrough in particle physics occurred. While Dirac's prediction of the existence of electron's mirror image had stirred the imagination of the theoretical scientists, no one expected any immediate discoveries associated with that. However, with no apparent connection to this prediction, in late 1930s, Millikan brought in a Post-Doctoral Associate, Carl D. Anderson, to build a large "cloud chamber" of $17 \times 17 \times 3$ cm size. Cloud chambers or Wilson chambers, invented by the Scottish physicist Thomas Rees Wilson, work with supersaturated vapors of liquids such as alcohol, in which particles leave tracks due to condensation along their ionizing path. Not only was the size of this particle detector large, Anderson also improved the system to drop the pressure in the chamber very fast to get the maximum supersaturation and improved sensitivity. The cloud chamber was placed, with one of the long dimensions along the vertical direction, in a powerful and uniform magnetic field of 2.4 T directed along the thin dimension. In such an arrangement, a charged particle traveling through the chamber would curve in the other long direction (perpendicular to the field) so that one would see the curving trail of the particle. The direction of the curve and the length of the track (range) determine the type of particle. An arc-light camera would be used to photograph the tracks.

In the summer of 1931, Anderson got his first images. Immediately he increased the maximum trackable energy of a particle like electron from 15 MeV to 5 GeV. After many such measurements, in the spring of 1932, Anderson found many tracks that looked similar to that of an electron, but indicated a positive charge with tracks bending the opposite way to the electron. The specific ionization, measured by counting the droplets along the track, showed that the positively charged particle also had unit charge as electron. At that time, the track was identified as that of proton, then the only known positive particle with unit charge. But there was a discrepancy. The electron paths accompanying these tracks corresponded to about 500 MeV and these tracks also had a similar but opposite curvature. If these were the heavier mass protons, that curvature would correspond to an extremely small energy (10–15 MeV). But the specific ionization should have been much larger for that type of proton (range would have been shorter) (Fig. 6.2). So they had to come to the radical conclusion that this might be an electron-like particle with a positive charge. Remembering that same particles with same energy would bend in opposite direction (motion direction is $\mathbf{v} \times \mathbf{B}$ – a cross product of velocity vector and magnetic field vector), Anderson wanted to rule out the possibility that these tracks were being made by electrons that came from below the table after some kind of scattering of the cosmic ray electrons. Such electrons coming from the opposite direction would curve the opposite way. Therefore, he inserted a 6-mm-thick lead piece in the middle of the chamber. The idea is that particles coming from above would leave a track of a certain radius of curvature above, but would slow down after passing through the plate, and therefore, the track under the plate would have a small radius of curvature, while the opposite would happen if the particle came from below. The result from that experiment unambiguously demonstrated that the new track was clearly not made by a scattered electron coming from below the chamber (Fig. 6.1). A particle entering the plate with a curvature and corresponding to a unit positive charge with energy of about 60 MeV had slowed down and changed to a radius of curvature corresponding to 23 MeV. A proton with that curvature would have had 20 keV after emerging and would have been absorbed quickly in about 5 mm length of chamber liquid, whereas the track was 50 mm long. Thus was obtained the "proof positive" for the existence of positively charged particle with electronic mass, postulated by Dirac. Later, Blackett and Occhialini confirmed this result with similar measurements but with an attendant Geiger counter which recorded the simultaneity of arrival of a cosmic ray particle. Later, many other sources and reactions would be identified for the production of these positive electrons.

The particle was named "positron" by Millikan and Anderson and in fact, to correspond to it, Anderson tried to name the electron, "negatron," but the name never stuck. Since this discovery, the presence of antimatter has been accepted and now the physics of antimatter is integrated into the so-called Standard Model, the most accepted theoretical model for fundamental particles. Since matter and anti-matter annihilate on each other to release energy in the form of photon, the reverse is also true, meaning that out of a gamma ray photon an electron–positron pair production can occur. (The question often asked is, how is it possible that a positron can travel through matter and leave a track without annihilating. This has to do with probabilities of reactions and even the annihilation process is not immediate. Typically, a positron has a nonzero lifetime of less than a nanosecond and, therefore, can leave a track of tens of centimeter in the cloud chamber (Fig. 6.2). However, what is noteworthy is that the positron track can never be much longer in a medium, while electron, depending upon its energy, can leave long tracks.).

In 1936, when Anderson was teaching a class as an Assistant Professor, Millikan broke into the room and told him breathlessly that Anderson had won the Nobel prize, which he would share with Victor Hess. Anderson is one of the youngest to receive the award. Even as he received this prize, speaking to King Gustav of Sweden in Swedish (his parents were Swedish immigrants to the USA), he was

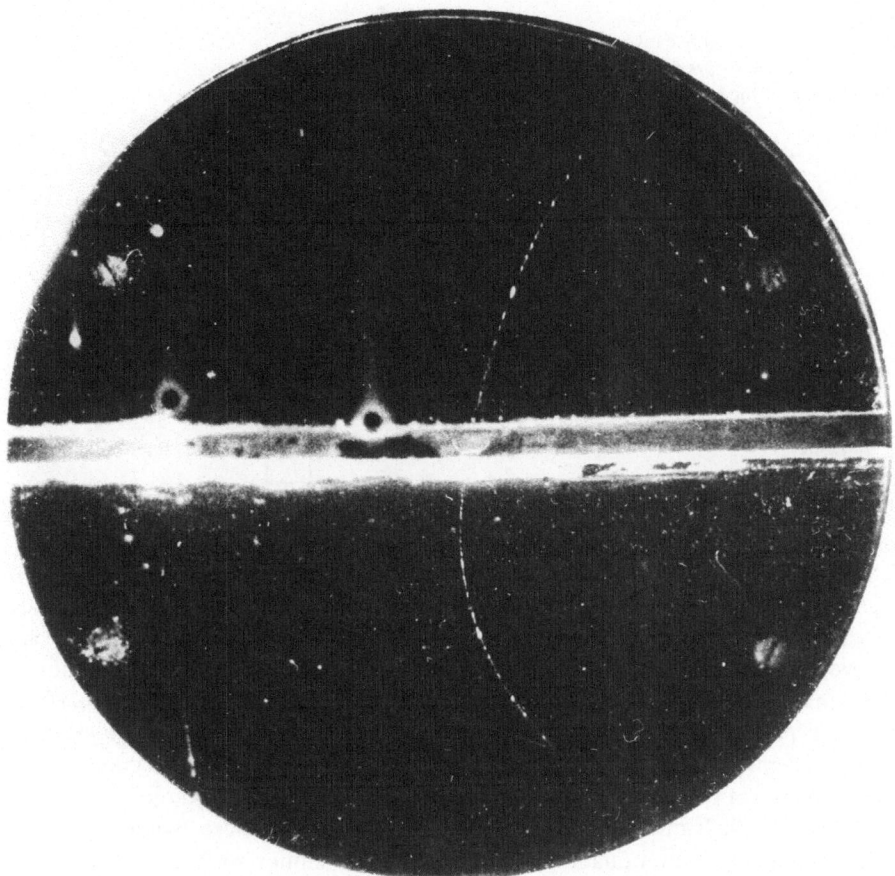

Fig. 6.1 A positron track, the direction of curvature and the change in radius of bending due to a magnetic field of 1.5 T after passing through the 6-mm lead plate from the bottom, showed that it is a positron (The positron came in with 63 MeV and lost 40 MeV by going through the lead plate). Image Ref. 10296273, Science Museum/Science and Society picture library, Carl D. Anderson, Physical Review, Vol.43, p. 491 (1933)

chalking up more particle discoveries to his credit. Once again, his discovery would be linked to another theoretical prediction, this time from the east.

The Art of Mesonry: Nuclear Force

As Chadwick discovered the neutron as a constituent of the nucleus, physicists started speculating on the nuclear force that was strong enough to bind the protons together despite their electromagnetic repulsion. They also speculated that what bound the proton to proton might also bind the neutron to the proton in the nucleus.

Fig. 6.2 The new particle could not be a proton because it was not easily stopped in the medium of the cloud chamber

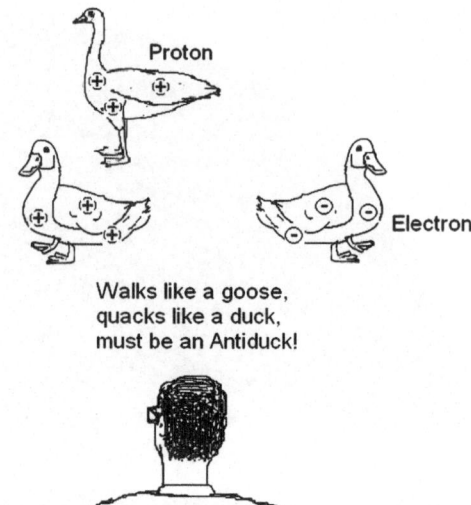

Proton

Electron

Walks like a goose,
quacks like a duck,
must be an Antiduck!

The solution that made sense of all this was proposed by the Hideki Yukawa, an Assistant Professor in Osaka University. Yukawa postulated a nuclear force, mediated by a new particle which is exchanged by the nucleons (protons and neutrons). This idea of an exchange particle to convey forces was borne out of a dilemma faced by Newton. The lack of a basis for action at a distance (force between particles without anything conveying it) had plagued Newton. But, by 1930s, scientists had understood that the electromagnetic force was conveyed by the exchange of photons, as if photons were bullets fired by the charged particles to push or pull each other. In quantum mechanical terms, these would be virtual photons existing within quantum mechanical uncertainties so that they can never be pinned down (detected experimentally). In the same fashion, in 1935, Yukawa, while working as an instructor, proposed another set of particles that would mediate the nuclear force binding nucleons together. His calculation of nuclear potentials (force is the negative of the gradient of the potential) agreed with the observations on nuclear interactions and even today, his theory is used for certain calculations. Later, it would be shown that this nuclear force is a net effect of the more fundamental "Strong Force."

Yukawa's theory was not an idle speculation. Since the nuclear force clearly stayed within very short bounds of the order of 10^{-15} m (1 F) compared to the atomic size of the order of 10^{-10} m (1 Å), most physicists thought that the nucleonic (attractive) forces must drop off quickly as $1/r^n$, where n would be around 7 (r being the distance). Thus, the force is confined within the nuclear shell. (The Coulomb force repelling an incoming positive particle that we encountered in Chap. 3 would be outside this shell.). The inverse square law for electromagnetic force due to a charged particle preserves the number of force carriers as the force emanates. This is because the photon carries the electromagnetic force. As the distance from the charged particle increases, the surface area of the sphere surrounding the charge increases as the square of the distance. This means that the surface density of photons decreases as the square of the distance. The fact that the force also decreases proportionally means that the total number of photons has remained the same. However, in the case of Yukawa force, we see that the nuclear force particle is too shy and is not seen too far outside the nucleus, and so its population must decay quickly (This is also consistent with a force that drops off fast.). Decays are conventionally of the form $\exp(-\lambda r)$, where λ is a constant related to the properties of the force carrier particle so that the nuclear force might vary as $\exp(-\lambda r)/r^2$ with distance, which allows for a fast exponential drop off of nuclear force outside the nucleus for $\lambda \sim 1/\text{nuclear radius}$. This form also explains the energies involved with nuclear binding. Yukawa defined the potential of the force as $\exp(-\lambda r)/r$, which does not quite give the form of force above, but the difference is negligible away from the nucleus. Further analysis shows that in such a potential, the force carrier must have a mass $m_U \sim \lambda h/(2\pi c)$, c being the speed of light and h, the plancks constant. The requirement that λ should be approximately the inverse of the nuclear distance leads to the mass of the force exchange of Yukawa particle of about 250 times that of the electron.

One other thing to note is that forces are supposed to be conveyed at about the speed of light. But this would mean that the time corresponding to this force conveyance would be about 10^{-15} m/3×10^8 m/s $\sim 3\times10^{-24}$ s. The lifetime of the Yukawa particle must also be of a similar magnitude, but when measured in the observer frame, relativistic effects would dilate the time by several orders of magnitude (by the relativistic factor $1/\sqrt{1-(v/c)^2}$, v being the speed of the Yukawa particle). The readjusted lifetime of about 0.01 μs agreed with observations on beta decay process (see later). The prediction of this particle also included the fact that there would at least be two kinds of these particles, one positive that would correspond to the change of proton to a neutron, and the other negative when a neutron changes to become a proton. In a very interesting result, the quantum mechanical description of the coupling between the meson (see below) and the nucleon gives the meson an electric charge that is identical in magnitude to that of the electron. This thoughtful rendering of the prediction was completely convincing and Yukawa's theory received much support from the theoretical developments in the USA. The hunt for such particles began. Since at the time there were no accelerators that could create mesons of that type of energy (billion electron volts), the search was made in cosmic rays.

Knock, Knock Who Is There? Meson. . .Meson Who?

Carl D. Anderson, discoverer of the positron, seemed to have once again made the first experimental strike in discovering a theoretically postulated particle. In 1936, he and Seth Neddermeyer burnt the midnight oil studying cosmic ray tracks in their large cloud chamber, and as they walked "up to Colorado Street and down to

Orange Grove, . . . and stop and have a glass of beer on the way," they talked about their results. They concluded that in their cosmic ray tracks, they had found a new negatively charged particle that had a strange mass – mass about 277 times that of an electron, much less than that of a proton. Anderson and Neddermeyer had not read Yukawa's paper because it was in Japanese. They named it mesoton (meso, Greek for intermediate – because the mass was intermediate between that of an electron and a proton), and Millikan, red in his face, ordered them to rename it mesotron, which sounds like electron. Though Anderson hated the name because it also sounded like cyclotron, a device not a particle, in their first paper in March 1937, they referred to it as mesotron. Later, the community would adopt the name "meson" at the suggestion of Werner Heisenberg. In early 1937, Jabez Street and Edward Stevenson at Harvard, who had also built a comparable large cloud chamber with a home-grown electromagnet, obtained tracks that also corresponded to the intermediate mass particles of positive and negative charges. (Some credit the discovery of this particle to Street and Stevenson). Soon after, Anderson and Neddermeyer also found the mesotron of positive charge. In 1937, Robert Oppenheimer and Robert Serber came to the conclusion that these must be the nuclear force particles proposed by Yukawa (some had called them Yukon).

In 1939, Y. Nishina and colleagues discovered particles in cosmic ray showers, with mass around 170–200 times the electron mass. It seemed that Yukawa's theory had been validated. All over the world, there was much activity on studying the properties of these particles, and over several years, experimentalists like Louis Leprince-Ringuet, Evan J. Williams, G. E. Roberts, and Bruno Rossi measured and poked to determine the properties and behavior of these new kids on the block. But progress and rigorous work were impeded by the World War and only after the war ended, focused attention was paid on the nature of these particles. There were serious discrepancies appearing between the particles discovered by Anderson and Neddermeyer and the other experimentalists' and Yukawa's particles. These particles did not interact with nuclei and went through matter like hot knife on butter, while the Yukawa particle should have been scattered strongly by nucleons (in the nucleus). Already in 1941, Tanikawa and Sakata suggested that the discovered particles of Anderson are not the Yukawa particle, but daughter products of decayed Yukawa mesons. In 1947, in a definitive study of these particles in Rome, a set of coincidence (multiple detectors measuring arrival of a particle at the same instant) measurements on cosmic rays were made by Oreste Piccioni, Marcello Conversi, and Ettore Pancini. They demonstrated that these particles with negative charge, even after being captured by a light atom, did not get absorbed by the nucleus immediately as is expected from a Yukawa particle, but stuck around and decayed into electrons in only a microsecond or so. But the predicted lifetime of a Yukawa particle was about 0.01 μs, and therefore, it was conclusively proved that Anderson's mesotrons were not the Yukawa particles. Confusion reigned until a few unlikely collaborators found the answer.

The above description simplifies the detection of mesons as if they all arrive intact in the cosmic ray showers. In reality, the mesons collide and interact with the atmosphere and produce a shower of secondary particles so that each meson particle produces a cone of secondary particles in three different types of cascades and few arrive intact. (First estimated by Homi Bhabha and Walter Heitler, 1936). Cloud chamber tracks had to be sorted out with an advanced knowledge of what was expected. This is an inexact process and it was, therefore, highly beneficial to detect these particles at a high altitude, where the cascade has not developed too much.

In early 1940s, Cecil Powell, son of a local gunsmith at Kent, England, worked at the University of Bristol in England studying ionization phenomena, building a Cockroft Walton generator and studying atomic scattering. But he found his love in a field of photographic detection of particles. As noted before, photographic plates were first used by Becquerel in 1912 to detect alpha particles. Since then, photographic plates were always used for particle detection, but they were not considered sensitive or highly usable compared to cloud chambers. The basic photographic plates contain an emulsion of silver bromide. In this emulsion, the lowest energy state is a lattice of ionized compound with positive silver ions and negative bromine ions. The surface of the plate (film) is preferentially occupied by silver ions. When a particle goes through the plate, an electron is knocked out of the negative bromine ion, which becomes neutral. Before the free electrons are captured again, they migrate to the surface and the positive silver ions cluster around the electrons and leave a track. This process requires a certain minimum threshold of energy transfer by the incoming particle, and simple emulsions do not meet the requirements for detecting particles. In 1937, Marietta Bleu pioneered the development of photographic, "nuclear" emulsions and caught nuclear integrations and resulting showers in a photographic plate.

Powell persisted in making nuclear emulsions more sensitive for detection. Powell and his team were proving that photographic plates, if coated with the correct nuclear emulsions, would be superior to cloud chambers for particle detection. A great advantage of photographic plates was also the fact that, like Hess's electroscope, they could be taken anywhere like mountain tops and on balloon flights. Scientists were using photographic plates supplied by Kodak to photograph the tracks left by particles in a cloud chamber and much of their frustration was due to their inability to resolve false tracks due to plate imperfections from the very few tracks possibly left by new particles. A new Ilford C2 emulsion reduced background fog, and several particles became visible in tracks that could not be seen in cloud chambers. In the meantime, Cesar Lattes, son of Italian immigrants in Brazil, was also working on photographic emulsions and had found that addition of Boron sensitized the emulsion greatly. He took some photographic plates to the top of Chacaltaya mountain, then the world's highest meteorological observatory in Bolivian Andes, and exposed them. In 1947, he took them back to work with Powell and Giuseppe Occhialini. Occhialini and Lattes also went up to Pic-du-Midi in Pyrenees in France and exposed plates spiked with borax. When they studied these plates, they found that these plates had rich and clear details. A young lady Marietta Kurtz first found a track that had unusual characteristics. While looking for more tracks like those, the group discovered new tracks of highly interacting heavier particles that agreed with all the predictions for the Yukawa particle. Lattes communicated a paper to Nature on the discovery without seeking permission from Powell. However, Powell was the one who received the Nobel Prize for the photographic emulsion development and the discovery of the Yukawa particle. To this day, Latin American scientists feel that it was an injustice that Lattes did not share the prize. In any case, the Yukawa particle was found and named pi-meson (or pion).

The other group that was disappointed in not sharing in the discovery was the Lawrence Radiation Lab which had built the 184-in. synchrocyclotron. So, it seems natural that Lattes went to work at the laboratory in Feb 1948 on Rockefeller Fellowship, armed with the photographic plates of the type he had used in cosmic rays. Almost immediately on his joining the team led by Eugene Gardner, Lattes identified artificially created pi-mesons (positive, negative, and neutral) by bombarding 380-MeV alpha particles (95 Mev/nucleon) on a target. They reported in Science that they obtained particle tracks on photographic plates placed near the target and these had the "same type of scattering and grain density with residual range found in cosmic rays." So, the first possibly fundamental particle had been created in a man-made accelerator. Once accomplished, this became the way. Hundreds of papers were then published on mesons and a series of mesons of were discovered, straining the Greek and English alphabets. In the 1920s to 1950s, like astronomers scrounging the skies for rare stars and planets, scientists had diligently looked for and caught the sparse drizzle of precious particles. But the cyclotron and the subsequent accelerators would provide a supply of particles and

Fig. 6.3 Untamed proliferation of mesons

some particles even became stored. Experiments with these energetic particles hitting fixed targets produced a parade of new particles, particularly mesons (today over 30 mesons are listed). Like the "Spaghetti Factory," "Cheesecake Factory," etc., particle factories would be born. The mesons were an unruly mob of particles (Fig. 6.3). Which, for more than a decade, frustrated and bewildered theoretical physicists, because they could not see how this much variety could exist (see *Nuclear and Particle Physics*, W.E. Burcham, M. Jobes Longman Scientific and Technical Publishing, Singapore (1995), for a table of Mesons). It did not sit well with the idea of a more orderly array of "fundamental" particles. These mesons were finally tamed only by the introduction of more fundamental particles called quarks by Murray Gell-Mann and George Zweig in 1964. These short-lived meson particles (the longest life expectancy of the pi-meson or the pion is 29 ns) are weirdly composed of both matter and antimatter. One meson (the B meson) even has an identity crisis and oscillates at a high frequency between two kinds of existences.

What of the particle discovered by Anderson and Neddermeyer, and Street and Stevenson in cosmic ray showers? It turned out that these were not nuclear particles at all, but were the equivalent of electron in a higher hierarchy of particles, rarely seen in normal matter. These are the muons which belong to the family of leptons. This particle is the one of the foundation stones for the Standard Model of particle physics. Muons do not interact with matter much and as a result, these are the most abundant cosmic ray particles at sea level losing about 2 GeV in the atmosphere and arriving at about 4 GeV. Muons decay in about 2.2 μs (long compared to the lifetime of mesons) in their own moving frame. Muons (antimuons) decay with the emission of electron (positron), a neutrino and an antineutrino.

The fact of the abounding variety of particles increased the need to push the boundaries of energy in order to study nuclear particles and to continue the search for the building blocks of nature. Bigger and better accelerators were needed.

Chapter 7
Rings of Earth: The Synchrotron

In the valley of the beautiful Appalachian mountains of Tennessee, USA, on the banks of Clinch river, just 25 miles north of the famous Tennessee Valley Authority dam, a sleepy farming community came together to retaliate against the Japanese attack on Pearl Harbor. In 1943, thousands of scientists and engineers descended on this community, now named OakRidge, to work on Uranium enrichment, which would lead to the Manhattan project. Sir Marcus Oliphant, Ernest Lawrence's trusted deputy, was one of them and was given the task of "watching the processes like an owl". However, he must have found time for other things, because that is when he came up with the idea of holding the particles in a single orbit as in the Betatron, but accelerating the particles with an alternating voltage in a cavity, as in Wideroe's concept, and wrote a memo to the Directorate of Atomic Energy, UK. In 1944, Vladimir Veksler of the Soviet Academy of Sciences in Moscow independently came up with a clearer idea, where the frequency of the voltage in the cavity would be varied, to take into account the relativistic reduction in orbital frequency at high particle energies. Before this paper was noticed in the West, Edwin McMillan of Lawrence Radiation Laboratory independently proposed a "synchrotron", in which the magnetic field would be ramped as the energy increased to keep orbit radius same, but also the frequency of the radiofrequency source would be varied. He used the word "phase stability" to describe the physics behind it.

The Betatron showed that it is possible to hold particles in a fixed orbit. The Betatron concept of ramping the magnetic field to keep the orbit constant while the energy increased was one aspect. As we saw, the ability to hold the particles stably in an orbit was also made possible only through the invention of "weak focusing". Holding the particles in a fixed orbit has many advantages both during acceleration and harnessing the high-energy particles. In all accelerators, the particles have to be held in as high a vacuum as possible because otherwise, they would be scattered by the gas molecules, atoms, and ions (charged atoms from which electrons have been stripped off). Such a scattering would also cause disastrous damage to the particle chamber, because the energy density (energy per unit volume) of this particle–gas soup could be high enough to destroy the chamber and even the magnets.

R. Jayakumar, *Particle Accelerators, Colliders, and the Story of High Energy Physics*, DOI 10.1007/978-3-642-22064-7_7, © Springer-Verlag Berlin Heidelberg 2012

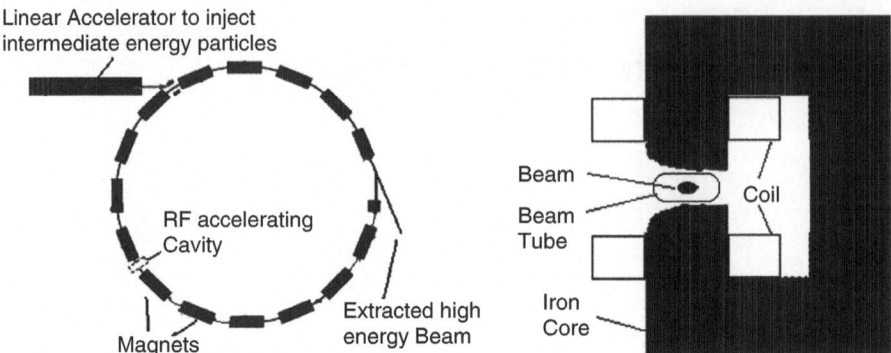

Fig. 7.1   A schematic of a synchrotron

High-Energy Physics Gets the Winning Ring from the Accelerator Physicists

If the particles could be held in a single orbit, a new concept takes hold. The chamber containing the particles would only have to be a so-called beam tube not larger than a few inches or centimeters (see Fig. 7.1). This, in turn, means that the magnets need not be hundreds of centimeters or inches in diameter. They would only have to provide the field over the diameter of the beam tube, which is a 100- to 1,000-fold reduction in the size of the magnet. This makes it feasible to obtain much larger magnetic fields required for much larger particle energies (For a given orbit radius, the required magnetic field is approximately proportional to the energy of relativistic particles and the energy the magnet requires is proportional to the square of the magnetic field strength and volume of the field region. Again, a reminder, once particles reach relativistic that is near-light speed, acceleration of particles does not increase their velocity much, but only their energy.) Alternatively, one has the option of reaching high particle energies with a limited magnetic field, but with a large orbit radius. The magnet weight increases only linearly and not as a square of the orbit radius, as in cyclotrons. One other advantage of a small beam tube is that the volume of the vacuum tube would also be small enough to permit easier pump down to ultrahigh vacuum.

Once this style of accelerator is chosen, the betatron acceleration with an inductive transformer is no longer preferable. For the betatron, as the orbit becomes larger for higher energies, the unused area enclosed by the orbit also becomes large. It becomes extremely wasteful to have to build a large magnet to fill this unused space with the alternating magnetic field. The magnet would become massive too. It is better to abandon the induced electric field concept and cause the acceleration every time as they pass through an RF cavity, as in Wideroe's linear accelerator. What is now the modern day application of this fixed-orbit accelerator, the accelerator and the magnet are decoupled, with the magnets doing the job of bending the path of the particles in a fixed orbit and keeping the particles focused in their path and a separate high voltage source with adjustable frequency accelerates the particle to higher energies.

Empowered by this concept of a small magnet and phase stability, the physicists were freed and indeed destined to increase particle energies continuously. Since the radius of the orbit is large, the particles were held between the jaws (poles) of a large number of small length magnets placed all along the orbit circumference (Fig. 7.1). The magnet cores enclosed a tube with vacuum in which the beam particles circulated. The initial synchrotrons had "weak focusing" magnets. The particles were accelerated to higher energy at one or two "posts" or radiofrequency cavities (resonating boxes) placed in the orbit. All iron synchrotron magnets were electromagnets, magnets powered by electric current in coils, placed around iron cores (Fig. 7.1). The iron poles were shaped to provide the "field index"(see Chap. 4) that is the spatial variation of magnetic field to get the required beam focusing in the vertical direction. The magnets did not have to be curved to the radius of the orbit (orbit radius was large) as long as the magnet length was small. Very importantly, the magnetic field was chosen independently of the resonance and betatron condition. The field would, however, be raised as in the betatron to keep the particles within the fixed beam line radius, as they gained energy. This meant that the radius of the "synchrotron" could be anything and the cost of the machine only increased, more or less, linearly with energy. In November 1946, D.E. Barnes and Frank Goward of Malvern Research Laboratory in UK demonstrated the first electron synchrotron by accelerating electrons from 4 to 8 MeV. In USA it was a General Electric team under John Blewett, which built the first synchrotron achieving 70 MeV. This machine produced the first visible synchrotron radiation. This was quickly followed by a 240 MeV machine built in 1948 at University of Rochester under Sidney Barnes. Two 300 MeV machines were completed in 1949 under Robert Wilson in Cornell University and McMillan at Lawrence Radiation Lab, who built a tight 1 m radius machine for accelerating electrons, which would have substantial synchrotron radiation. It is said that this machine had such a powerful magnet and made such a tremendous noise that people imagined that the noise was due to powerful electrons smashing into targets. Around the world, more Synchrotrons were quickly built and achieving higher particle energies would be only a matter of effort and money.

Synchrotron, Pride of Nations

1947 – the year in which the cold war officially started with the Truman Doctrine, State of Israel was created, India and Pakistan became independent, first airplane broke the sound of speed, the first documented sighting of UFO took place, Jackie Robinson became the first African American to be permitted to join the US major baseball league, and transistor was invented, was also a watershed year in the history of science, particularly physics research. Across the globe, it became clear that with the rapid pace of scientific and technical developments, particles could be accelerated to phenomenal energies. From the fundamental and theoretical physics side, there was breathless anticipation of the unveiling of new physics, related to nuclear forces and radioactive decay processes. Theory was taking fast strides and experiments with GeV (10^9 eV) energies were needed to verify many theoretical ideas. Simple back of the envelop calculations showed that such machines would be great machines, which could not possibly be built by individual Universities and large centralized research facilities had to be established to house and operate these massive accelerators with the collaboration of the Universities. This conclusion spurred the birth of several national and international laboratories.

Brookhaven National Laboratory and Cosmotron

In 1946, Universities of Columbia, Cornell, Harvard, Johns Hopkins, Princeton, Pennsylvania, Rochester, and Yale and Massachusetts Institute of Technology formed a nonprofit corporation which would run a large accelerator research facility. The facility, Brookhaven National Laboratory, was established on March 21, 1947, in an old army base, under the supervision of the newly formed U.S. Atomic Energy Commission (AEC), which also provided the funding for the initial start up and for the construction of the accelerator. The BNL is now one of the world's best high-energy physics laboratories, now housing the Relativistic Heavy Ion Collider (RHIC).

Once formed, there was an intense debate about the first machine the Laboratory should undertake. Stanley Livingston, yes, the same physicist, who had build the first accelerator to cross the 1 MeV barrier in the (Lawrence) Berkeley Radiation lab and the head of BNL's project, preferred a safer, more reliable, and cheaper 284 in. (about 7 m), 600 MeV synchrocyclotron (see Chap. 5), while I.I. Rabi, in Columbia University, wanted the laboratory to go aggressively after the plum – a 10 GeV synchrotron. He exhorted his fellow physicists to "be a little wild" and go for the 10 BeV (billion eV). [Making Physics – A Biography of Brookhaven National Laboratory, 1946–1972, Robert P. Crease, The University of Chicago Press, Chicago (1999)]. Rabi kept the pressure on and the 10 GeV idea was accepted more broadly. AEC too was persuaded by Rabi's arguments and decided to pursue this machine. Livingston, who was on loan to BNL, was disheartened by this choice and left the laboratory 6 months later to return to MIT, but continued to contribute. Thus was born the first large GeV class synchrotron called the Cosmotron and demonstrated the propensity of physicists to gamble occasionally.

The AEC had decided that the newly formed BNL, east coast laboratory, and (Lawrence) Radiation lab in Berkeley each would build a GeV class machine, but could afford to build only one 10 GeV machine. Leland Haworth led BNL to wisely choose the more conservative 3 GeV machine, but with a promise from the AEC that BNL would get to build a much larger machine later. The name chosen for the 3 GeV synchrotron was Cosmitron (representing an ambition to produce cosmic rays), but was changed to Cosmotron to sound like cyclotron. The beam size of

64 × 15 cm (25″ × 4″) and an energy goal of about 3 GeV determined the machine parameters. In a laboratory, which functioned in the style of University departments with loose management with no up-down hierarchy (see Making Physics, Robert Crease), invention was present at all levels, cutting corners, cutting costs, and implementing somewhat risky design trade-offs. As per the idea of John Blewett, the magnets were designed to be C shaped as opposed to the traditional H shape. As seen later, this fateful decision would lead to one of the most important discoveries in accelerator design.

The synchrotron consisted of 288 magnets each weighing 6 tons and providing up to 1.5 T, forming four curved sections of 30′ (about 9.1 m) radius separated by 2′–10′ (about 3 m) long straight sections to inject the beam and, for the first time in accelerator history, extract the beam. Each magnet had a gap of 9″ (22.5 cm), a pole length of about 3′ (about 90 cm) and an iron cross-sectional size of 8′ × 8′ (240 × 240 cm). The magnetic field decreased radially by about 6% over the pole face to ensure vertical stability of the particle orbits. The magnetic field was ramped up during acceleration first in proportion to the velocity (orbit radius is proportional to the particle velocity) and once the particles start acquiring relativistic speeds where now the mass increases, the field was increased in proportion to the energy increase (orbit radius proportional to Lorentz factor γ of the particle). The range of field change was kept within limits by first accelerating particles to an intermediate energy in another accelerator and then injected into the Cosmotron. To adjust to the change in the orbit frequency as the particle (proton) became relativistic (mass increases), the radiofrequency of the accelerating cavity would have to change by a factor of 13. All these presented a terrifying new regime of design and fabrication. To accommodate the large range of frequencies, the first use of new materials called Ferrites (first manufactured by Philips) was made. (Ferrites are magnetic materials fabricated using powder metallurgy techniques and have low losses in alternating current devices and can also give somewhat higher magnetic fields.)

During construction, doom was constantly predicted as, one by one, physicists discovered serious obstacles in different areas – eddy (image) currents in wall beam tube, beam instabilities because of noisy component in the radiofrequency source, and so on. Worse, Berkeley's mock-up experiments indicated that the Cosmotron beam would blow up. The straight sections without magnets were worrisome, because there was no focusing and the betatron oscillations would change suddenly and might swing wildly. But, all these major problems were overcome. The machine had many interesting new features besides the first use of ferrites. In 1950, Gary Cottingham invented the "quadrupole" magnet to focus the beam in order to transport it from the primary beam source of a Van de Graf generator; major waste of power was avoided when during each 1 s of charging the magnet coils, a 800 kW generator driven by a flywheel was used, and during the discharge cycle, the coils drove the generator now working as a motor and revved the flywheels back up.

In March 1952, the construction and installation of the Cosmotron (Fig. 7.2) were completed and team members found that the beam disappeared in about third of a second. Even though this is the longest any beam had been obtained, this was

Fig. 7.2 The Cosmotron at the BNL (Courtesy Brookhaven National Laboratory History Images, Image CN4-427-49)

still disappointing. Someone even wryly suggested that maybe protons live only that long. After 3 months of work, the beam disappearance was found to be due to a sensor that was faulty and was interfering with the beam. One can only wonder how many foreheads were slapped and whose head rolled. On 20th May 1952, the crew turned on the machine, obtained 200 and then 400 MeV, and when they came back reenergized by lunch, they reached 1 GeV practically immediately. It is no wonder that this was sensational news and journalists proclaimed "man-made cosmic rays" were here. Popular Science magazine called it the biggest sling shot in the world. In June, the machine reached the benchmark energy of 2.3 GeV. With additional coils on the pole faces, the Cosmotron reached the peak energy of 3.3 GeV in January 1953. In 1960, the machine accelerated 100 times the number of particles (of the order of 10^{10}), compared to the first beam. At the dedication of the machine, the AEC commissioner declared that Cosmotron would serve for the "enlightenment and benefit of mankind". At this point, the Cosmotron had cost $8.7 million, a record for a physics experiment. The Cosmotron worked until 1969. During its operation, it was a great source of cosmic ray-like particles and helped resolve one of the greatest mysteries of nature, namely, conservation of parity (More about this later.)

The Bevatron

There are many stories about the collaboration and competition between Brookhaven National Laboratory, Lawrence's Berkeley Radiation Lab, and the emerging powerhouse of Europe, CERN. As stated earlier, in doling out the construction money, the Berkeley lab was awarded the 10 GeV machine, the Bevatron. Contrary to the approach taken by BNL, this accelerator was designed and fabricated like a tank using very conservative estimates. Sobered by the first tests on beam instabilities, the beam tube was chosen to be $4' \times 14'$ (1.2×4.2 m) and the joke was that, if built, it would be the first machine to accelerate jeeps. But when the Cosmotron succeeded with its aggressively small beam tube, the Lawrence lab's Bevatron beam tube was reduced to a sane $1' \times 4'$ size, a 15-fold reduction. The calculation of the beam envelope taking into account betatron oscillations was very hard and the Cosmotron and the later machines provided good benchmarks for what was achievable.

Declaring that the main purpose of the 40 m (130 ft) size Bevatron, with about 10,000 tons of magnets, was to produce antiprotons, Lawrence reduced the peak energy to 6 GeV, the threshold for production of antiprotons, which proved correct. Bevatron started operation in 1954 and achieved 6.2 GeV in 1.85 s with a beam that did not deviate more than a few inches and behaved impeccably. With the success of this machine, more synchrotrons followed, for example the 12.5 GeV Zero Gradient Synchrotron (ZGS) at the Argonne National Laboratory near Chicago, USA, which was established in early 1960. The largest of them all is the Dubna Synchrotron at 10 GeV energy with a radius of 28 m and with a weight of the magnet iron of 36,000 tons and is still in operation. So, this basic concept of holding the particles in a fixed orbit with magnet segments and a separate accelerating module is here to stay. With this concept, modularization of the components such as magnets becomes possible. Miles of a chain of such modules can be built to construct accelerators of large radius which would be required for very large particle energies when cost and technology limit the strength of the bending field. (Another basic physical limitation is that the radius of the orbit cannot be reduced too much since at high energies and small orbit radii, synchrotron radiation would become a large energy loss, limiting the acceleration.) The synchrotron concept made it possible to make a breakthrough in the GeV class, enabling new spectacular physics discoveries leading to the coming of age of the Standard Model for particle and progress toward the Grand Unification Theories.

Catching the Antiproton Traveling Faster Than the Speed of Light

Lawrence's dedication of the Bevatron to the discovery of antiproton was not an empty dream or wishful thinking. The 6 GeV energy was the best estimate of the beam energy required to create the antiproton, by collision of a beam proton with a target nucleus. In a collision with a stationary target, a particle cannot release (give up) all its energy, because the net momentum has to be conserved and the particles coming out of the reaction must have the momentum of the incoming proton. Antiprotons can be created only along with another proton (the incoming proton collides with a target nucleus proton or a neutron and gives up some of its energy and from that energy, a proton–antiproton pair is produced and they carry the momentum). The incoming proton has to create two particles each with a mass of proton (equivalent rest energy = 0.938 MeV each). But the pair must also have the kinetic energy corresponding to the conservation of momentum of the beam proton creating the pair. The most copious production of antiprotons was calculated to be for the antiproton velocity of 87% of speed of light or a Lorentz factor $\gamma = 1/\sqrt{(1 - (v/c)^2)} = 1/\sqrt{(1 - (0.87)^2)} = 2.03$, so that the energy of the antiproton $= \gamma \times$ rest energy $= 2.03 \times 0.938 = 1.9$ GeV. The energy required to create the pair is a minimum of 2×1.9 GeV $= 3.8$ GeV. But the particles coming out of the target collision have to conserve the momentum. Since the incoming particle is a proton and the outgoing particles are three particles, all with proton mass (two protons and one antiproton), kinematics and momentum conservation require that they share the incoming proton momentum equally after the reaction. Therefore, the beam energy required for the best chance of creating antiprotons is $3 \times 1.9 = 5.7$ GeV. Accurate calculation gives a value of 5.6 GeV. Lawrence set the Bevatron energy just above it to 6 GeV and the machine actually reached 6.2 GeV.

The cast of characters (Fig. 7.3) who would vindicate him were the cream of the crop in their intelligence, experience, and tenacity. Emilio Segre', first student of the grand physicist Enrico Fermi, had discovered a new element Technetium, had found that plutonium could be used for building nuclear bombs, and contributed greatly to the understanding of nuclear forces. Segre' was a bright intellectual, who wore multiple hats in UC Berkeley as well as in the Los Alamos National Laboratory. He was not shy to flex his managerial muscle as well to get the resources to the team he managed. Owen Chamberlain, an Associate Professor with a quietly seething but crystal clear mind, was a favorite of the UC Berkeley students. Clyde Wiegand, a student of Segre' and one who was born with a technical prowess to devise, invent, and build, was the third. Ed Lofgren, Wilson Powell and their team were the other cast members. The Bevatron was, of course, the star of the cast. Chamberlain heard that Ed McMillan and Maurice Goldhaber, accomplished physicists, had bet that antiprotons did not exist. After all, no one had seen them in cosmic ray observations. Chamberlain decided "By Jove, this is what I want to do." (meaning, he wanted to find the antiproton). Chamberlain teamed with Wiegand to devise an apparatus that would measure the charge and mass of particles coming from a target, hit by the proton beam from the Bevatron. In order to measure the expected mass (same as proton), they chose to measure the momentum and velocity, while first selecting only particles whose paths were bent in a magnetic field, in the direction consistent with a negative charge. (Antiproton has negative charge.) Lofgren's team would look for the star pattern of tracks left in a photographic emulsion (see previous chapter on positron detection), due to annihilation of the antiproton with a proton. Powell's team would look for it in cloud chamber. Lawrence ran the Bevatron like a space station experiment, with researchers getting assigned dates for their experiment. It was the luck of the draw how well the Bevatron worked on any given day. Sometimes people would trade their days because they had or did not have the right equipment and calculations. One such trade favored Chamberlain's team over Lofgren's and they discovered the antiproton first, though Lofgren could easily have been the first.

Chamberlain and Wiegand chose the parameters and instruments with simplicity, but great diligence. First they chose their instruments after talking to many experienced experimenters. Following Oresto Piccioni's suggestion, they selected a spectrometer, in which the antiproton would have a curved trajectory because of the application of the magnetic field. Since the field is accurately measured, the radius of the curvature would be known for a given velocity (radius $= \gamma mv/qB$, with m and q being the mass and charge of the antiproton). They chose a velocity of 77% of the speed of light for catching a particular track. With the radius known, the particle stream coming out of the magnet was selected for this radius by a slit and only antiprotons with that velocity would go through it. This selected the desired momentum and charge. They had two scintillation detectors (detectors that emit light when particles traverse them), spaced 11 m apart. If it is the antiproton with the right velocity that is going through them, then scintillation time of the two detectors would be apart by 50 ns. This identified the particle velocity. Then they had the Cerenkov detector ("pickle barrel" shaped "Secret Weapon"), which was built following the design of Sam Lindenbaum. Chamberlain had the insight to know that this design was selected for velocity of the particle while most thought the Cerenkov detectors give signals only for velocities above a certain value.

A word about Cerenkov detectors (these would be further described in a later chapter): These detectors are based on a discovery by Pavel Cerenkov who found that when particles travel faster than the speed of light, they emit a radiation. What, you ask? Travel faster than the speed of light? But that is forbidden by relativity, you say. Actually, not exactly. Matter cannot travel faster than speed of light in vacuum. But in a medium such as water, light slows down considerably. For example, speed of light in water is only 75% of the speed in vacuum. So, the antiproton that Chamberlain chose would exceed this speed and the Cerenkov radiation would be emitted. This is what Chamberlain and Wiegand set out to measure, except that they used a slab of glass as the medium but instead of using it as a "threshold" detector which would emit light for any particle with a speed greater than the speed of light in glass, they used the additional information that the angle of emission depends on the actual velocity. The light due to a slower superluminal particle would be emitted at a shallow angle and faster particle would result in a larger angle. They, once again, selected the angle for a particle with 77% of the speed of light. With these gauntlets, any signal they got had to be due to an antiproton corresponding to the expected velocity. It would be a lucky break that would lead Chamberlain's team to be the first to discover the antiproton. Six weeks of operation of Bevatron had been scheduled in August and September 1955. Six days into the operation, the machine broke down and when it was again operational on September 21st, it was Lofgren's team's turn to do the experiments. Instead, Lofgren loaned his team's time to Segre' and Chamberlain. That would turn out to be a big gift. Sure enough, they found the unmistakable signals of the

Fig. 7.3 *Left* – The team that discovered the antiproton, Emilio Segre, Clyde Wiegand, Edward Lofgren, Owen Chamberlain and Thomas Ypsilantis, (Courtesy: Lawrence Berekeley National Laboratory Image Library, Image XBD9606-02969.TIF), *Right* – Edward McMillan and Edward Lofgren standing at the top of the radiation shielded Bevatron, (Courtesy: Lawrence Berekeley National Laboratory Image Library, Image XBD9705-02170.TIF)

antiproton! Soon, the other two teams also discovered the evidence of the antiprotons. As in the case of such discoveries, no parades were thrown for our heroes, but there was tremendous celebration for confirming a major aspect of physics and its implications were and are enormous. Unfortunately, the Berkeley Daily Gazette reporter misunderstood the implication, reporting a "Grim New Find", because he thought he came close to being blown up by the antiproton-reporter collision-annihilation. On a side note, Segre' reported the discovery to the Vatican, but happily, unlike Galileo, he was not tried by the church and found "vehemently suspect of heresy".

The Yin and the Yang of Accelerator – Alternating Gradient Synchrotrons

Chinese philosophy holds that alternating flux of Yin and the Yang forms the balance of the Universe. The gain, growth, and advance of the one mean the loss, decline, and retreat of the other. This principle is embedded in a new way of focusing and equipped with this, the modern day accelerator builders are scanning the very horizons of achievable energy.

Builders of weak focusing synchrotrons recognized that such synchrotrons would be limited in energy because the stability (focusing) of the particles in the horizontal direction is marginal. An assured stability in the vertical direction was needed for reasons of cost and size. The pole gaps had to be minimized to reduce the magnet and beam size, which also meant that the beam had to be well focused in the vertical direction and higher the field index the better is the vertical focusing. But this reduces the focusing in the radial (horizontal) direction. In the case of Cosmotron, already the field index n (which is a measure how fast the field decreased horizontally) had been increased to very close to 1 at the energy of

3 GeV, due to the saturation of iron, and there was no margin to raise the particle energy any further. (As stated in Chap. 5 on Betatron, the field index, proportional to the gradient of the magnetic field, had to be maintained between the values of 0 and 1 to obtain orbit stability in both vertical and horizontal radial directions). This orbit instability in the horizontal plan is accentuated at higher energies and would entail a large beam pipe. Only other way to improve focusing would be to increase the strength of the magnetic field. Magnet currents would then have to be increased substantially to obtain even somewhat higher field, due to the magnetic saturation of iron. Bigger beam pipes or stronger magnets would mean larger and exorbitantly expensive magnets to reach high particle energies.

It would be serendipity that would give a major breakthrough to overcome this obstacle and would change all accelerator designs forever. It was noted above that Livingston and John Blewitt's team had, in a cost and space saving measure, chosen C-shaped magnets in the Cosmotron, as opposed to the conventional H-shaped magnets used in cyclotrons and synchrocyclotrons (See Fig. 7.4). Now, in the Cosmotron, the flux return yoke (vertical leg) of the C was on the inside of the orbit (circle). This made it easy to get at the beam from outside the orbit and to extract negatively charged "secondary" particles, created by beam hitting a target. But, as can be seen readily, unlike the H shape, the C magnet shape is unsymmetric and produces different magnetic field gradient on the inside vs outside. On the inside (leg side) of the pole face, the field reduces less rapidly than on the outside and one could not get magnetic field gradient which was symmetric with respect to the centerline of the pole faces. This was not a problem at low magnetic fields. As the magnet field is increased for increasing particle energy, the iron core saturates. The return leg leaks out more and more field lines into the air and the magnetic field, in the space between the pole faces and the return leg, increases. The field would then decrease more strongly toward the open pole face. This has the effect of increasing the negative gradient in the magnetic field and the field index would become greater than 1, as the magnetic field was raised and the iron got saturated.

Focus at Last: Lemonade out of Lemon

In 1952, during his summer visit to BNL, Livingston had a novel idea. He wondered whether he could alternate the orientation of the C-shaped magnets, meaning that the return leg of one magnet would be inside of the orbit and the neighboring magnet in the ring would have its vertical leg on the outside of the orbit (see Fig. 7.5). For lower magnetic fields and low particle energies, this would not change anything since the iron saturation does not kick in. But at higher energies and fields – the vertical magnetic field decreases with radius strongly (larger n) and the particle would be well focused in the vertical direction, but would be defocused in the horizontal radial direction with increasing amplitude of radial betatron oscillation. But in the reversed orientation with the iron leg on the outside of the orbit, the vertical magnetic field would increase with radius and this would focus the particles radially (horizontally) and defocus the particles vertically.

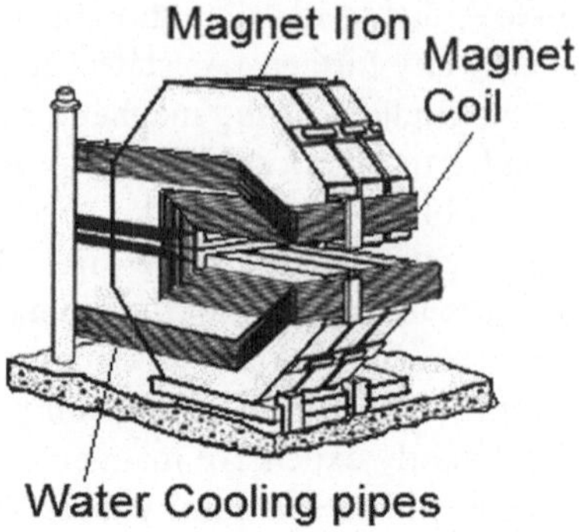
Magnet Iron
Magnet Coil
Water Cooling pipes

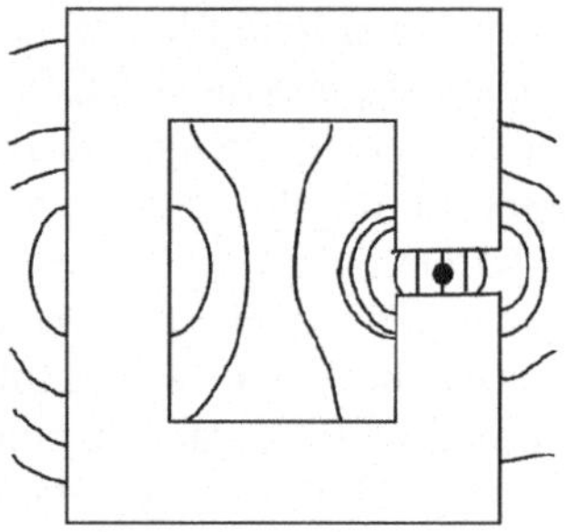

Fig. 7.4 *Top* – The Cosmotron magnet (Courtesy: Brookhaven National Laboratory), *Bottom* – Nonsymmetry of magnetic field lines in such a C-shaped magnet. For clarity, the crowded flux lines in the iron and the coil are not shown

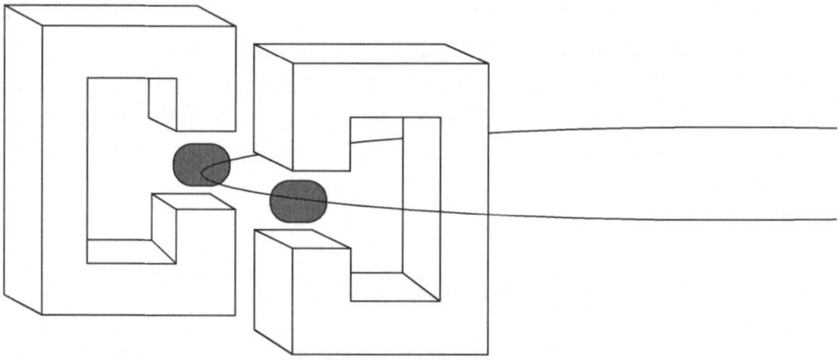

Fig. 7.5 Alternating arrangement of C-shaped magnets. The *shaded area* shows the beam tube region. The center of the ring arrangement is toward the right; that is, the radial direction is toward the left. In the first magnet (*right*), the gradient is strongly negative (field decreases with increasing radius), and in the second (*left*), gradient is strongly positive (the field increases strongly with increasing radius)

He reasoned that over the particle orbit, this would average the gradients and a desirable field index value of $n = 0.6$ (as stated before n has to be between 0 and 1) could be achieved in both inward and outward radial directions. Livingston was torn by the attractiveness of the alternating arrangement, but was very concerned about the orbit stability at high fields (high energies) with this type of arrangement and was also worried that, in the new arrangement, secondary electrons created by interaction of beam particles with the walls would not be swept out as was the case with an ideal field index. After vainly trying to estimate the effect, on a fated day, he decided to ask the Ernest Courant, the well-known Columbia University physicist-mathematician, for help. Courant was not very encouraging at first. But he knew precisely how to set up the mathematics of this problem and immediately checked it out.

Courant was astonished with the result. As Livingston had suggested, indeed, on average this arrangement kept the beam intact and contained in both the directions. But what surprised him was that this arrangement was much better than just arranging the field gradient in a narrow regime for only weak focusing and gave a much more focused beam with smaller betatron oscillations. Next day, with a quizzical expression, he conveyed this discovery to Livingstone and Hartland Snyder; it was evident that they had struck gold. They could see that the beam envelope could be even smaller than what it was. Snyder immediately remembered the analogy with optics, where a light beam could be kept collimated with a succession of focusing (convex) and defocusing (concave) lenses (Fig. 7.6). This analogy is almost exact and this type of optical ray tracing calculation is carried out even today for beam simulation. (There is a small subtlety that needs to be noted; actually, the arrangement is more focusing than it seems because, for a given incident angle at the lens, the focusing lens provides more focusing than the defocusing lens defocuses). This also corresponds to an intuitive understanding of how a ball can be kept up in air by allowing it to fall and then hitting it up. There is a so-called Earnshaw Theorem proved in 1842 that particles (in that case) cannot be focused in two transverse directions simultaneously, because potential fields do not have a minimum to which particles can move to. So the Alternating Gradient alternates the focusing in two directions, focusing in one direction while letting go in another.

Livingston had suggested some moderate field indices (gradients) for checking this out. But, if these modest gradients are good, would stronger gradients be better? So Livingston suggested a value of n (field index, which is proportional to the field gradient) audaciously larger than the usual field index of less than 1, n of -10 and $+10$! Yes, it was better! Then even higher. It then became clear that this was not a matter of index – the larger the index (even 1,000), the tighter the beam (smaller the cross section) – this was just a new principle of focusing accelerating beam of

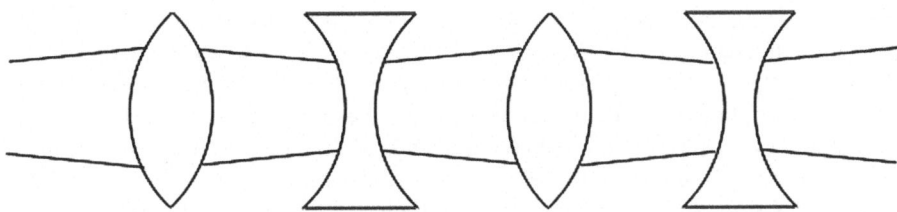

Fig. 7.6 Alternating convex and concave lenses for keeping a beam collimated within an envelope

energetic particles! So, it became evident that the beam could race through to very high energies, on the stepping-stones of these "lenses" while being confined to a beam of small radius. A small beam means small vacuum tube, with small magnet pole gap, which then considerably reduces the energy for powering the magnet and the overall size of the magnet for a given energy. Two papers were sent on this important invention and scientists and engineers got excited about new possibilities. They even dreamed of 30 GeV on the Bevatron being constructed in Berkeley, but Blewett got the approval for such a machine (named Alternating Gradient Synchrotron – AGS) to be constructed in BNL itself. In a separate paper, Blewett pointed out the applicability of AG focusing to linear accelerators (see Chap. 8). The invention of AG focusing changed everything. Beams that were centimeters wide would now be millimeters and microns wide. After this, there were truly no fundamental physics limits to achieving energetic particles.

Incidentally, Livingston and even Courant give credit to an invention of vertical focusing by L.H. Thomas in 1938 and say that their "alternate gradient (AG) focusing" was anticipated by Thomas. M.K. Craddock has shown that this is not true and Thomas focusing is a very different prescription for cyclotrons. (Paper presented at 23rd Particle Accelerator Conference, Vancouver, Canada, TR-PP-09, May 2009.) However, the invention had indeed been anticipated and there is an interesting human-interest story on this invention. The invention had already been made by Nicholas C. Christofilos, a U.S. citizen of Greek extraction who had been stranded in Greece during World War II and had taught himself physics from books distributed by the Germans. Born in Boston, MA, to Greek parents, he moved to Athens and worked there as an electrical engineer in an elevator company. In the 1940s, after his interest in radios and transmission was curtailed by Greek prohibitions during the Nazi era, Christofilos attended to repairing army trucks. Finding a lot of time on his hands, he applied himself to learning physics through German books. After the war, Christofilos started his own elevator firm, but by then he had become a knowledgeable and inventive physicist in his own right. In 1946, he invented the synchrotron all by himself, but of course, by then it had already been invented by others.

His patent rights had been recognized and patent was awarded in 1956. In 1949, he wrote to University of California Radiation Laboratory (UCRL) at Berkeley describing his independent invention of the Alternating Gradient Focusing which he called strong focusing [with the term strong referring to the fact that the oscillation wavelength is small compared to the orbit (ring) circumference]. The Berkeley lab did not look at it and just filed it away. When the invention of the same principle, independently by Courant, Livingston, and Snyder, was hailed as one of the biggest discoveries, Christofilos was still working in his elevator firm in Greece. Christofilos happened to come across the Physical Review article describing the details of this concept. He immediately recognized that this was his own concept of "strong focusing" and naturally believed that someone had stolen his idea from his letter to UCRL. As soon as he could, he flew to USA, staked his claim on the discovery, and showed his patent to prove it. When he met with the scientists and members of the AEC, they admitted his claim to the invention and paid him a handsome fee of $10,000 for the use of his idea. Eventually, Christofilos joined the BNL team and made several other contributions to science, including the development of a device called Astron at the Lawrence Livermore Lab, to confine thermonuclear plasmas.

Alternating Gradient: Strong Focusing Synchrotrons

The first alternating gradient synchrotron accelerating electrons to 1.5 GeV was completed in 1954 in Cornell University, Ithaca, NY. Other strong focusing synchrotrons were also built at this time, for example, the Tokyo University (1.5 GeV), 6 GeV machines in Hamburg, and at MIT in Cambridge, MA, USA. This latter machine had special magnet laminations and would ramp up at 60 Hz rate with a very high frequency RF! In 1955, when the University of Hamburg offered the Chair of Physics to the famous Viennese physicist and a visionary Willibald Jantschke, he asked for DM 7.5 million, an audacious demand, to build a state-of-the-art accelerator laboratory. Thus was born the Deutches Elektronen Synchrotron (DESY) with 7 GeV final energy. Another noteworthy machine was the 4.5 GeV NINA at Daresbury in Liverpool, England. The Brookhaven laboratory's Alternating Gradient Synchrotron (AGS), occupying the equivalent area of six football grounds, reached 33 GeV in 1960. It had magnets constructed so as to simultaneously produce a bending magnetic field of 1.15 T (for a beam

Fig. 7.7 CERN's 28 GeV Alternating Gradient Proton Synchrotron. The alternating gradient can be seen as one C magnet with the open side facing out and the next facing in, as envisioned by Livingston. Courtesy: CERN, European Nuclear Research Organization, Image Reference : CERN-AC-5901157)

bending radius of 85 m) and a field gradient of 4.5 T/m. This accelerator was a source of several major discoveries. The discovery of the theoretically postulated omega-minus particle in the famous 80 in. bubble chamber detector (see later) in 1963 gave the first step toward getting the zoo of particles tamed according to an order and led to the Standard Model of particle physics. At this time, a quantum chromodynamics (QCD) model had been developed to explain the nuclear force that binds particles such as protons and neutrons and mesons to each other, through the "strong force" between the more fundamental constituent particles called quarks. In QCD, a property that is ascribed to a quark is termed as "flavor". In 1974, AGS beam was used with a new 7 ft bubble chamber to discover the "charmed" (one of the flavors) quark, proving the model. The discovery of the muon-neutrino, Charge-Parity (CP) violation, and the discovery of J particle, in the AGS, each brought Nobel prizes.

While the accelerator-synchrotrons have changed in many of the concepts, electron synchrotrons continue to be built for the purpose of generating synchrotron radiation in the X-ray wavelength region and beams and radiation from these are staple of nanodevices and semiconductor industry.

Phoenix Rises Out of the Ashes of War: Emergence of CERN and Construction of the Proton Synchrotron

Europe (including Great Britain) had always dominated the field of physics discoveries till the late 1920s and even in the early 1930s. Discovery of early concepts in science to discovery of electrons to theory of relativity to founding quantum mechanics were feathers in the caps of the nations of Europe. But the two world wars not only destroyed lives, but also the infrastructure and memory of many institutions. The scientific institutions were in disarray either because they had been recruited into war efforts or because scientists fled the countries to escape the Nazi onslaught. When the World War II ended, there was much regret about this past, but also the vehement spirit to regenerate the scientific institutions and spirit that Europe had engendered and fostered. Europe saw the writing on the wall that science would be the engine of technological inventions and economic opportunities. Despite the past long and glorious history of each of the illustrious laboratories in Europe, these institutions could not take on the task individually. Fortunately for Europe in the late 1940s, it had been realized that future particle (high-energy) facilities, with novel and sophisticated instruments and machines, would have to be constructed under an umbrella of a large laboratory. In 1949, Louis de Broglie led with the proposal at the Cultural Conference in Lausanne, Switzerland, to create a European research center. Scientists like Niels Bohr, Isidor Rabi, Pierre Auger, Edoardo Amaldi, Raoul Dautry, and Denis de Rougemont took up the cause immediately in a spirit that heralded national and international cooperation. In June 1950, at a UNESCO conference in Florence, Italy, Rabi placed a resolution authorizing UNESCO to "assist and encourage the formation of regional research laboratories in order to increase international scientific collaboration. . ." In February 1952, 11 countries signed an agreement establishing the European Council for Nuclear Research (CERN). At the Council's third session in October 1952, Geneva was chosen as the site of the particle physics laboratory – a glorious beginning for a series of glorious scientific developments to take place in future. After the formal ratification by individual governments, CERN came into existence in September 1954.

In June 1952, CERN adopted the proposal put forward by Heisenberg to first build the relatively safe and straight 600 MeV synchrocyclotron (Fig. 7.8) (fondly known as SC) and also start work on a more ambitious 10 GeV weak focusing synchrotron. Construction of the 600 MeV SC was started in 1955, applying all the lessons learnt in the previous machines and with a conservative design. Unsurprisingly, the machine achieved its target energy immediately after commissioning in 1957. This was the beehive of the scientific type bees for many years and rare decays and beta decays of meson were observed and quantified and contributed to the understanding of muon physics.

In summer of 1952, even before a site was chosen for the center, a CERN team led by Odd Dahl would visit the Brookhaven National Lab (Fig. 7.9). Odd Dahl was this Norwegian adventurer with no formal physics experience, but had taught himself Physics. At the age of 24, he had been chosen to pilot a plane on a polar expedition. But while taking off from ice, the plane broke up and he had to spend an

Fig. 7.8 CERN's first accelerator, the 600 MeV synchrocyclotron. Courtesy: CERN, European Organization for Nuclear Research, Image Reference: CERN-HI-7505285)

Fig. 7.9 CERN visitors to BNL – (*left* to *right*) George Collins (Chairman, Cosmotron group), Odd Dahl, Rolf Wideroe (then at Brown Boveri), and Frank Goward from CERN. (The quark machines: *How Europe fought the particle physics war* By Gordon Fraser, Institute of Physics, London (1997))

astounding couple of years on the ice. He used his time to develop and construct intricate instruments and made geophysical observations around that ice-locked Siberian region. He then went on to be part of a team building a Van de Graaf generator at the Carnegie institution in USA. He also had plenty of experience as a research physicist, having worked at the Norwegian Dutch Reactor. Dahl had been invited by Pierre Auger to help in the preparatory work in establishing the CERN facility. CERN had expressed its wishes to be modeled after the Brookhaven collaboration and Heisenberg had stated that the Synchrotron (now named Proton Synchrotron – PS) was to be built like the Cosmotron, except that it would be scaled up to 10–20 GeV. Little did any one know that the CERN visitors would return with a happy bundle of gifts. Naturally, when they learned of the upcoming visit, Blewett, Livingston, and all the other physicists were excited and nervous and wanted to make a good impression. In an effort to make the Cosmotron look really good, Livingston started thinking of ways to improve Cosmotron's performance. This is what led to the idea of alternating the orientation of magnets and the quick succession of dominoes falling to enable the invention of the alternating gradient concept. When the CERN team arrived in the first week of August 1952, they were expecting a dull recital of Cosmotron features, the organizational arrangements, introduction of various personnel, and some nice lunches and dinners. Instead, they found breathless physicists barely suppressing their excitement over the revolution-ary idea of "alternate gradient focusing". Though the US AEC scolded the scientists for releasing the information that might have military implications, the American physicists shared this exciting news with the CERN group. At once, the European scientists boldly recommended the adoption of this, as yet, untested concept for its new synchrotron and the energy was raised to 30 GeV. The council showed its audacity in accepting this proposal and the 30 GeV Proton Synchrotron was born. In spite of building competing machines, which were amazing in their scope for that time (BNL was building the 33 GeV AGS – Alternating Gradient Synchrotron), the two laboratories collaborated closely. John and Hildred Blewett from BNL were part of the team building the machine in CERN.

In 1956, a conference on the PS (proton Synchrotron) was held and included surprise guests from the Soviet Union, who not only supplied great ideas, but also provided precious Vodka, brought in after valiant struggles with Swiss customs official. The design was finalized. The PS machine, with over 628 m in circumference, would have C-shaped magnets with alternating orientation arranged to great precision (within a few cm) required for the strong focusing machine (Fig. 7.7). The best engineers from around the world joined in the efforts to obtain this precision in alignment. The ground was broken to start construction in May 1954 and the machine was fully commissioned on 16th September 1959 and J.B. Adams thrilled the attendees of the Accelerator Conference in Geneva that the proton beam had gone around a full circle. It would seem that the PS had beaten the AGS in Brookhaven by 6 months or so. But after that the machine just did not look right. For a few days there were bits of exciting news, but overall, it seemed that the beam was being lost as the accelerating field was turned on. Hildred Blewett, Brookhaven physicist and a collaborator at CERN states (A night to remember – CERN Courier, October 30, 2009), states, *"but most of the time we had been discouraged, puzzled by the beam's behaviour, frustrated by faulty equipment or, after quick trials of this remedy or that, in despair over the lack of success. The protons just didn't want to be accelerated."* The PS group was getting demoralized with no window into what was going on. Part of the problem could be attributed to the fact that exciting experiments were going on in the SC (the 600 MeV Synchrocyclotron) and the installation and testing of various subsystems were hampered because the accelerator physicists were too busy with the SC. This meant that the accelerator could be tested with beam only on 2 days a week.

In the accelerator system, the beam actually spirals with betatron oscillations and would suffer some radial drift due to conditions being not accurately correct. A feedback system was put in place, where a sensor would measure the radial shift of the particle and accelerating radiofrequency source would be automatically adjusted in attempt to correct this radial shift of the beam. In early October, they tried this. To their delight, this improved the life of the beam, but alas, at about 2–3 GeV energy the beam was lost again. The matter stood there and the physicists were losing heart. Even cold cut meats, cheese, and bread in the control room could not induce them to come on time for the beam start-up on operational days.

On 24th November, Hildred Blewett was considering returning to Brookhaven, because the AGS needed her. After a listless dinner, John Adams and Blewett walked back to the PS control room, with Adams apologizing to Blewett for wasting her time in PS, while she could have been working on the AGS. Blewett mentioned that Wulfgang Schnell had been trying something new and that might work, but she could see in Adams' eyes that he did not believe it. But as they reached the control room, they had a surprise. In Blewett's words (CERN courier, Oct 30, 2009),

"...Schnell had gone ahead over the last couple of weeks wiring it up for a quick test. Just a few days before, I had been down in the basement lab, listening to his enthusiasm. The idea was to use the radial-position signal from the beam to control the r.f. phase instead of the amplitude. With this system, the sign of the phase had to be reversed at transition and, in his haste, Schnell had built this part into a Nescafe tin, the only thing of the right size, available at hand.

Adams opened the door to the Central Building. For a moment the lights blinded us, then we saw Schmelzer, Geibel and Rosset – they were smiling. Schnell walked towards us and, without a word, pulled us over to the scope. We looked... there was a broad green trace... What's the timing... why, why the beam is out to transition energy? I said it out loud – "TRANSITION!"

Just then a voice came from the Main Control Room. It was Hine, sounding a bit sharp (he was running himself ragged, as usual, and more frustrated than anyone), "Have you people some programme for tonight, what are you planning to do? I want to...". Schnell interrupted, "Have you looked at the beam? Go and look at the scope." A long silence... then, very quietly, Hereward's voice, "Are you going to try to go through transition tonight?" But Schnell was already behind the racks with his Nescafe tin, Geibel was out in front checking that the wires went to the right places, not the usual wrong ones. Quickly, quickly, it was ready. But the timing had to be set right. Set it at the calculated value... look at the scope... yes, there's a little beam through... turn the timing knob (Schnell says that I yelled this at him, I don't remember)... timing changed, little by little ... the green band gets longer... no losses. Is it... look again... we're through... YES, WE'RE THROUGH TRANSITION!". On the second try, "....with the blessed phase-control and the Nescafe tin. Change timing on the scopes, watch them and hold your breath. One second (time for acceleration) is a long time. The green band of beam starts across the scope... steadily, no losses... to transition... through it... "

Why were they so excited? They were excited because this meant that the beam had crossed the 10 GeV, energy at which the beam's response to the accelerating voltage changes abruptly through what is known as gamma transition. What is this transition? Time for another accelerator physics titbit:

Longitudinal Focusing (Phase focusing) and Gamma Transition

In a particle beam, all particles are not borne equal and there is a spread of beam energies and therefore beam orbit radii. When accelerating the beam, if the accelerating voltage is not at the right phase when the different particle arrives, then the beam would be lost because the spread would increase and particles would separate from their bunch. There is a very nice solution for shepherding these particles. If there is an RF source with an oscillating frequency equal to the correct frequency corresponding to the design γ (relativistic energy factor), then it can be phased in the following fashion for a proton beam: When the slower proton arrives, the RF source should present a negative voltage so that the slower proton gets accelerated a little extra and it should present a positive voltage to the faster proton so that it is slowed down (see Fig. 7.10). This is essentially the idea that Schnell was

Fig. 7.10 The waveform of
the RF phase and the
appropriate phase at which a
faster and a slower particle
should be arranged to arrive

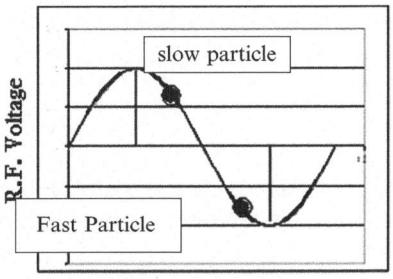

Phase of R.F. Voltage

using with his Nescafe tin apparatus. The phase of the RF control waveform was
tuned to work this way. Since B is increased along with the beam energy (γ), the
nominal (central) frequency remains the same and the particle orbit radius also
remains the same.

But, there is a complication. In the relativistic regime, the orbital frequency is
dependent on the particle speed. The orbital frequency of a particle with a rest mass
m, velocity v, and charge q is given by $\omega = qB/(\gamma m)$, where B is the bending
magnetic field and $\gamma = \sqrt{(1 - (v/c)^2)}$, c being the speed of light. We also saw that
the nominal orbit radius $r = (\gamma m v)/(qB) = p/(qB)$, since $p = \gamma m v =$ the nominal
momentum. Now if a proton in the beam bunch has a slightly higher momentum,
$p + \Delta p$, then the orbit radius of this particle would be $r + \Delta r = (p + \Delta p)/(q
(B + \Delta B))$, where the errant proton also sees a slightly different magnetic field
because of its different radius and timing. From this relation we can calculate the
fractional change in the orbit radius for a given fractional change in the proton
momentum. The ratio

$$\alpha = \frac{\frac{\Delta r}{r}}{\frac{\Delta p}{p}} \tag{7.1}$$

is called the momentum compaction factor. Now the time taken to go around the
orbit (say circular orbit) of radius r is

$$T = \frac{2\pi r}{v} \tag{7.2}$$

(We have to keep track of the velocity, even though it is close to speed of light.)
So the time taken to go around by our errant proton,

$$T + \Delta T = \frac{2\pi(r + \Delta r)}{v + \Delta v}, \tag{7.3}$$

$$\Delta T/T \sim \left(\frac{\Delta r}{r}\right) - \left(\frac{\Delta v}{v}\right) \tag{7.4}$$

for $\Delta T \ll T, \Delta r \ll r, \Delta v \ll v$. Algebraic manipulation of the expressions $p = \gamma m v$, $\gamma = \sqrt{1 - (v/c)^2}$, and derivatives give,

$$(\Delta v/v) = (1/\gamma^2)(\Delta p/p). \tag{7.5}$$

This shows that as particle energy and γ increase, velocity increase due to momentum (energy) increase becomes less and less. Substituting

$$\frac{\Delta T}{T} = \left(\alpha - \frac{1}{\gamma^2} \right) \frac{\Delta p}{p} = \left(\frac{1}{\gamma_t^2} - \frac{1}{\gamma^2} \right) \frac{\Delta p}{p},$$
$$\gamma_t = \sqrt{\frac{1}{\alpha}}, \tag{7.6}$$

where γ_t is the so-called transition γ. So, that is the quirk: when $\gamma < \gamma_t$, that is, $1/\gamma^2 > 1/\gamma_t^2$, $(\Delta T/T)$ is negative for positive $(\Delta p/p)$; that is the time taken by our errant proton is less if it has a larger momentum (as would be expected) and vice versa. When the protons have accelerated to a γ larger than γ_t, $(\Delta T/T)$ is positive for positive $(\Delta p/p)$; that is the time taken by our errant proton is longer if it has a larger momentum and vice versa.

Basically, there are two competing factors: one is the reduction of time period due to increase of velocity and the second is the increase in time period (decrease in orbital frequency and increase in orbital radius) as the energy (γ) increases. In the first regime, the velocity increases quickly with energy while γ increases only slowly, and in the second regime, the reverse happens. One gets a transition from one regime to the other at $\gamma = \gamma_t$, the transition gamma. Most synchrotrons worry about going through this transition smoothly, because there is a great chance of losing the beam, if the control of the phase of the RF is not changed correctly at this transition. Schnell had arranged to automatically feedback any radial excursion of the beam to the RF phase adjusting network and this took care of the longitudinal focusing of the beam so that they are all constantly kept at the same orbit radius, independent of which side of the transition it was. (There is an undesirable consequence for the acceleration process once gamma transition is crossed. The time period around the orbit increases and like the ions in the betatron, the rate of energy gain in a given time period decreases, for a constant rate of ramping of the magnetic field; but fortunately the dependence of time period on gamma becomes weak at high gammas).

The transition once crossed, the CERN PS proton beam kept going and, in the very same try, reached 25 GeV. The incredulous scientists saw that the beam had 10^{10} protons, ten times more than they ever expected. While Hines pored over any and all meters to eliminate any doubts, the news spread like a fire and revelers kept pouring into the control room. There were unsolicited drinks of gin and vodka, which magically appeared to celebrate the occasion and scientists started to discuss the experiments they were itching to do. Hearing the news in Brookhaven, John Blewett heaved a sigh of relief and satisfaction (smug?) because the AGS builders had already decided to use a phase control feedback arrangement.

Table 7.1 Comparison of the AGS and PS parameters (Advances in electronics and electron physics, by L. Marton, Vol 50, p.74, Academic Press, NY (1980))

Machine parameter	AGS (BNL)	PS (CERN)
Machine radius	128 m (421 ft)	100 m (328 ft)
Injection energy (MeV)	50	50
Final energy (GeV)	32	26
Phase transition energy (GeV)	7	5
Aperture (vertical × horizontal)	7.6 × 15.2 cm (3″×6″)	8 × 12 cm (3.15″ × 4.72″)
Number of cells	12	10
Number of gradient reversals	120	100
Field index	360	282

Above is the comparison of the AGS and the PS to show how similar these machines were (Table 7.1). The PS after having served physics independently, will now, in its mature age, contribute as an injector to the LHC for the next 25 years.

With PS (and also AGS) functioning well in the coming years, the accelerator physicists had learnt longitudinal and transverse dynamics of high-energy, high-intensity beams. They had learnt how to build these machines with tight mechanical tolerances and what sophisticated RF and electronic systems were required. This would stand in good stead for the future grand machines. The PS would go on to achieve 10^{13} protons per pulse, the very next year, a record for this class of machines. The PS machine provided the beam for the Nobel Prize winning discovery of weak force neutral currents. The accelerator physics and technology had matured with these machines and even today this is the basic scheme in circular large accelerators, except for the use of quadrupole magnets for focusing.

Quadrupole Magnets: The Four-Legged Particle Sheep Dogs

The original idea for the strong or alternating gradient focusing was to combine the magnetic field gradient (magnetic field decreasing or increasing radially within the beam tube) and the bending magnetic field in the same magnet. But, the same paper that reported the invention by Courant, Livingstone, and John Blewett at BNL had also suggested the use of so-called quadrupole magnets for linear accelerators that did not require bending magnets (for focusing only the field gradient is important). The quadrupole magnet had been invented by Gary Cottingham of BNL in 1950. In a quadrupole magnet shown in Fig. 7.11, there are two north and two south poles, instead of the one north and one south pole in dipole bending magnets. This generates a cusp at the center of the magnet with zero magnetic field at the axis (center). The magnetic field increases linearly outward from the axis toward the poles, providing a gradient. In the new scheme of synchrotron with quadrupole, the orientation of successive magnets is alternated to give the alternating gradient. The figure below shows the field configuration. In a conventional iron magnet (Fig. 7.12), the north and south would be pole pieces, usually made of iron, magnetized by copper windings.

Fig. 7.11   Quadrupole
magnetic fields, *left* (F) –
focusing vertically
defocusing horizontally, *right*
(D) – defocusing (vertically),
focusing horizontally

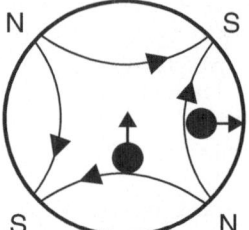 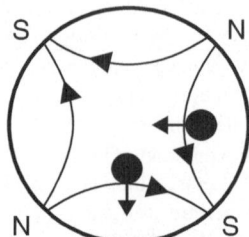

Fig. 7.12   A quadrupole magnets in the Japan Proton Accelerator Research Complex (Photo credit: KEK/Japan Atomic Energy Agency J-PARC Center)

When conventional (normal conductor) windings are used, the maximum gradient that can typically be $1/a$ T/m, where a is the radius of gradient region in meters. So, for $a = 10$ cm (0.1 m), the maximum gradient attainable is about 10 T/m. The present day synchrotrons do not use the field index based magnets and use separate bending (dipole) and focusing quadrupole magnets. The use of these quadrupole magnets separates the bending function and focusing functions and accelerator and magnet designs and requirements become simpler and modular. A typical cell of a synchrotron consists of "FODO" cells, F for focusing quadrupole, which focuses in the vertical direction and defocuses in the horizontal (see Fig. 11.7), O for bending magnets, and D for defocusing quadrupole. The first large accelerator to pioneer this design was the Fermi National Laboratory in Batavia, IL, near Chicago.

When we think of these quadrupole as lenses, the focal length of a quadrupole lens, with a gradient B' and length l, as seen by a particle with charge q and momentum p (relativistic mass x velocity), is given by $F = p/(qB'l)$. For example,

30 T/m quadrupoles with a magnet length of 1.7 m would have a focal length of approximately 25 cm for an energy of 400 GeV. (Therefore, as the particle is accelerated, both the bending and focusing magnetic fields have to be increased in proportion to the momentum, to keep the particles in the same radius and with the same focused (betatron) conditions.)

Accelerator Beam Dynamics

The maturity of accelerator physics is closely linked to the complete understanding of beam behavior in the presence of magnetic fields and RF fields. Now, in designing an accelerator, the particle beam response to the accelerating electric field and the periodic magnetic structures – bending and focusing, etc., is theoretically calculated and suitably matched to meet the goals of the accelerator. With a well-constructed machine, these goals can be met.

Longitudinal Dynamics

When a particle arrives at an RF cavity, it accelerates or decelerates depending upon the sign and magnitude of the alternating voltage at the instant it arrives; that is, the energy gain or loss depends on the arrival phase of the RF field in the cavity. Since the intent is to accelerate the particle and yet keep the orbit radius of the particle constant by suitably increasing the bending and focusing magnetic fields, the energy gain, the RF voltage, and the rate of increase of magnetic field are linked for synchronism. In this, a synchronous particle (a particle with the target energy/ momentum) will stay synchronous (arrives at the same correct phase at the accelerating structure). For a particle with a mass m, charge q, and velocity v corresponding to the relativistic factor γ, the radius of the orbit is $r = p/(qB)$, where B is the bending magnetic field and p, the particle momentum $= \gamma m v$. Therefore, with $p = qrB$, the magnet has to be ramped according to the rate of change of momentum with time.

$$\mathrm{d}p/\mathrm{d}t = qr\mathrm{d}B/\mathrm{d}t, \tag{7.7}$$

where $\mathrm{d}B/\mathrm{d}t$ is the rate of change of the magnetic field. If the particle arrives at the RF cavity at the synchronous phase ϕ_s, then the accelerating voltage is $V\sin\phi_s$ and the particle would gain an energy $\mathrm{d}E_s = qV\sin\phi_s$, where V is the peak voltage of the RF source. The rate of change of the particle momentum $\mathrm{d}p/\mathrm{d}t = (1/v)\mathrm{d}E_s/\mathrm{d}t = \mathrm{d}E_s/\mathrm{d}z$, since $z = vt$ is the distance along the synchrotron ring. Taking the whole ring with a total circumferential length of L_R,

$$\mathrm{d}p/\mathrm{d}t = qV \sin \phi_s/L_R = qr\mathrm{d}B/\mathrm{d}t \text{ (above)} \tag{7.8}$$

or

$$V \sin \phi_s = L_R r dB/dt, \tag{7.9}$$

which gives the synchronous condition linking parameters of the synchrotron. In slow ramping synchrotrons, dB/dt and therefore $V\sin\phi_s$ are kept constant.

But, particles have a spread in energy and many would be nonsynchronous to various degrees. A nonsynchronous particle that arrives when the RF voltage is at a phase ϕ_n acquires a nonsynchronous (extra) energy, $dE_n = qV(\sin\phi_n - \sin\phi_s)$. In that process, it advances by a phase $d\phi_n$ proportional to dE_n after each turn in the synchrotron, so

$$d\phi_n = CqV(\sin\phi_n - \sin\phi_s). \tag{7.10}$$

This equation is an oscillatory equation for the phase of the particle and the particle exercises oscillatory advances and retreats in phase around the synchronous phase (gets ahead and behind the RF voltage oscillation) with an angular frequency

$$\omega_n = 2\pi\nu_n = \sqrt{(CqV \sin\phi_s)} \tag{7.11}$$

gaining less or more energy than the synchronous particle on each trip. (This oscillation is now in the orbital direction and is different from and in addition to the betatron oscillation. The longitudinal oscillation is not a back and forth displacement but is a going ahead and lagging behind type of oscillation). The frequency is called the synchrotron tune and is of the order of 0.001; that is, the beam makes one full oscillation after going a several hundred turns. If the machine is designed well, the particle will stay in a bounded phase range and will have an overall stable longitudinal oscillation about the synchronous phase.

Betatron (Transverse) Dynamics

As stated in Chap. 5, a particle oscillates transverse to the beam tube as it goes around in an orbit in a magnetic field with a gradient. When generalized to a Strong focusing FODO cell (see above), these betatron oscillations are still like simple harmonic motions, but represented by the well-known equation of motion, called the Hills equation

$$\frac{\partial^2 u}{\partial t^2} + k(z)u = 0, \tag{7.12}$$

where u is the transverse direction (x or y) and z is the beam (longitudinal – going around) direction. The difference between this equation and an equation for a spring is that here the spring constant k is not a constant and is a function of z. Since the beam radius is much smaller than the ring radius, a cylindrical approximation works well enough. The solution of this equation is of the form

$$u = \sqrt{\varepsilon\beta(z)}\cos(\psi(z) - \psi_0),\qquad(7.13)$$

where β is called the betatron function, ε is the emittance, which is a constant for a beam, $\psi(z)$ is the phase of the oscillation, which is only dependent on the variation of β, such that the local wavelength is equal to $2\pi\beta$, and ψ_0 is a reference phase. This expression states that particles will oscillate in transverse (x and y) directions and their displacement at any particular location depends on the emittance, the betatron function, and the initial phase of each particle. Since both the amplitude and the phase of the oscillation are functions of the location, the oscillations would look different from a sinusoidal wave with a constant k. If β (minimum) $= \beta^*$, then the betatron function varies nominally as $\beta = \beta^* + z^2/\beta^*$. But within a FODO cell, it rises and falls (see Fig. 7.13; note that this is not displacement, but the actual particle displacement is given by u.)

While the particles do this dance around the beam axis in the transverse direction, bending of the particles can spray the particles and disperse them. The so-called chromatic effect or particles having a spread of energy is the reason. When there is a spread in energy, different particles would be bent slightly differently causing this dispersion. (This is analogous to dispersion and separation of white light into its components, or a water hose spraying water when the water stream falls to the ground.) The strong focusing FODO lattice also controls this "Dispersion Function", which is approximately the dispersive displacement for a unit error in the momentum ($u_d = D\Delta p/p$). Like the betatron function, alternate gradient focusing causes the dispersion function also to be periodic (see figure).

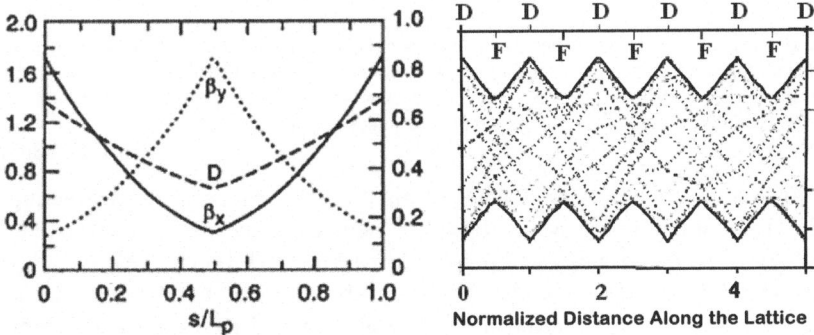

Fig. 7.13 (*Left*) Example of betatron functions (x direction and y direction-left scale) and dispersion function (dashed-right scale); (*Right*) The actual paths of particles with different momenta and phases being contained within an envelop, as the particle traverses several FODO cells. The horizontal axis is for orbital distance (along the synchrotron) and the vertical axis is the transverse displacement (arbitrary units)

Given all this, one can see that a cell and indeed the whole accelerator can be precisely designed based on the calculated betatron motion, RF phasing, and choice of β^*. As the beam goes through each FODO cell, it advances its phase by an angle $\psi = 2\sin^{-1}(L/2F)$, where L is the spacing between the quadrupoles and F is the focal length of the quadrupole (see previous chapter). Therefore, if there are n FODO cells going around the synchrotron, the total phase advance is $n\psi$, and therefore, the "betatron tune" or the number of oscillations the beam makes going around is given by

$$n\psi/2\pi = n/(\pi)\sin^{-1}(L/2F).$$

Orbit stability requires that L/F be positive and be less than 1. For the Tevatron, with a focal length of 25 cm, and inter-quadrupole length of 30 cm, the betatron tune is approximately 20 (corresponding to strong focusing). (Actually, an integer tune means repeated pass over the same path by a particle and errors will add and there can be no stable orbits. Therefore the tune is always made to be a noninteger). As one can see, the beam particles oscillate in the transverse direction (go around in a spiral motion) much more often than in the longitudinal oscillation with a synchrotron tune of only 0.001 or so.

The other parameter entering the (7.13) is the emittance and it is a very important parameter describing how narrow is the spatial spread and spread of the particle velocities. This parameter, as we shall see later, determines intensity of particle collisions in a collider. In an accelerator, the emittance remains constant around the ring and only very special conditions can change the emittance of the beam.

The "Main Ring" and the "Main Man"

> "It has only to do with the respect with which we regard one another, the dignity of men, our love of culture. It has to do with: Are we good painters, good sculptors, great poets? I mean all the things we really venerate in our country and are patriotic about. It has nothing to do directly with defending our country except to make it worth defending."
>
> Robert R. Wilson

In 1969, in the congressional hearing on the justification for constructing the multimillion-dollar Fermi Main Ring, US Senator John Pestore questioned Paul McDaniel of Atomic Energy and persisted in asking its military "defense" application. Robert Wilson, the would-be Director of the National Accelerator Laboratory, sitting next to McDaniel, asked if he may answer and answered thus.[Fermilab: physics, the frontier, and megascience, By Lillian Hoddeson, Adrienne W. Kolb, Catherine Westfall, University of Chicago press, Chicago, NY (2008)].

The successful completion of the AGS in Brookhaven and the PS in CERN had energized the community and made them greedy for say a 300 GeV machine. So the Midwestern Universities Research Association (MURA), based out of University of

Wisconsin, Madison, proposed the so-called Fixed Field Alternating Gradient machine, capable of achieving high beam intensities, as the cheapest. Mathew Sands, an "iconoclastic" (see Lillian Hoddeson, "Beginnings of Fermilab", Golden Books, fermilab History and Archives project (1992)) physicist from California Institute of Technology (Caltech), could not resist accepting the challenge and proposed an accelerator based on a daisy chain (a cascade) of multiple accelerators. The cost and complexity of an accelerator are higher if the particle has to be injected at a low velocity and extracted at high energy. This is related to the fact that the beam envelope is large at low energy and gets smaller at higher energies and one has to design the machine with a large beam tube aperture which is not needed at high energies. But at high energies, space is a premium and one would want a small beam tube. In Sands' proposal, the energy range in each accelerator would be limited; a 10 GeV booster machine would accelerate and inject multiple pulses of beam into the main accelerator. A team consisting of Sands, Alvin Tollestrup, Robert Walker, Ernest Courant, Snyder, and Hildred Blewett estimated the cost of a 300 GeV machine as $77 million, truly a bargain. This design was focused on scalability and reduced cost. A Western Accelerator Group (WAG) consisting of Caltech, Universities of California Los Angeles and San Diego, and University of Southern California was formed to propose the design, but the Berkeley Radiation Lab team now under the leadership of McMillan did not join and would propose its own design. A BNL proposal was more expensive – $300 million for a 300 GeV machine.

In 1961, while 30 physicists gathered to pore over design choices, the community and the AEC were wringing their hands on the politics of fairness and cost. The broad physics community had discussed the importance of the discovery of two types of neutrinos (ghostly particles which are now part of the final count of fundamental particles), were anxious that the Strong Force and Weak interactions and the relationship between these forces were poorly understood, and were on the side of a high-intensity, multi-100 GeV accelerator. The Berkeley design was in the middle-cost class of about $150 million and Berkeley deserved the next machine, since the AGS – Alternating Gradient Synchrotron – had been built in BNL. But the WAG machine was the cheapest. As for the location, while each group including MURA campaigned for their home locations, WAG understood that it had no chance of winning it, against such powerful competitors. The competition on this aspect was becoming a feud and the US AEC decided to take charge of the project and the "vibes" until 1963. Paul McDaniel of AEC instigated the formation of a High Energy Accelerator Advisory Committee, now renamed the High Energy Physics Advisory Panel (HEPAP), to provide a forum for the discussions. Even as this committee was meeting, the AEC decided to form a Presidential Advisory Committee under Norman Ramsey, a Harvard Professor known for his accomplishments in physics and with a reputation of being fair and diplomatic. At this time, the Berkeley Radiation Lab (with the cooperation of the WAG) put forward a more parochial and expensive proposal for a 100–200 GeV machine at $152–$263 million. This cost included all experimental facilities and stated that the project would be built and managed by Berkeley.

After many heated debates and memos, in 1963, Ramsey panel recommended a 200 GeV machine of the Berkeley design, a 10 GeV electron synchrotron at Cornell University, and start of studies for a 1 TeV class of machine at Brookhaven. (This would become the superconducting Isabelle at BNL). The panel even as it recommended a second priority 12.5 GeV machine under MURA, on the condition that the higher energy machine should be a higher priority, it knew there would be trouble. After all, MURA had been working for a decade to create a high-energy physics laboratory in the midwest America. But the panel decision was essentially a death sentence for the MURA project. Ramsey's recommendation of the parochially managed machine at Berkeley Radiation Lab was quietly resented and opposed by leading physicists, who feared that under the traditionally parochial and top-down hierarchy of the lab's culture since 1930s, a national character of the facility would not be established. To further fuel this suspicion, McMillan, then Director of the Radiation Lab, held fast to the Radiation Lab's sole management, despite AEC admonishments.

Fortunately for the high-energy physics community, luminaries, and Nobel Laureates like Enrico Fermi, I.I. Rabi, Oppenheimer, and Robert Bacher were vociferous and convincing proponents of funding for research and were a uniting force. In 1962, Cuban missile crisis had intensified the cold war and in policy circles, physics advancement was becoming a strategic necessity. People started seeing scientific achievements as American achievements. Therefore, the physics community knew that the society valued them and their work. Several discussions and reviews later, young Leon Lederman, already an icon among physicists, in an erudite and humorous talk, proposed a new lab coining the term TNL – the Truly National Lab, intended equally for all national users and one in which outside users would have equal access. (This was also a dig at the BNL, which had started shedding its Cosmotron style of democratic functioning.) At the dedication of SLAC, President Johnson had emphasized "on time" and "within budget", which have now become litmus tests for the success of scientific projects. This type of pressure and recognition of the valuable suggestions, such as the Booster staged accelerator from Caltech, made the Radiation Lab realize that it would be worthwhile to collaborate on the facility. As a result, a 10 member national team was appointed by the AEC to oversee the design and a 3,700 acre site was proposed in the Livermore valley in California.

Middle America Stands Up

Meanwhile, the MURA group, having realized that it had lost their machine in the race anyway, decided to vie for locating the machine in the midwest. At this time, the Argonne National Lab's 12.5 GeV machine had been overshadowed by the AGS. Bernard Waldman, the MURA Director, and Elvis Stahr, President of Indiana University, stirred up the debate and used phrases such as "step child" and "Midwest does not get a fair shake," etc. Midwestern politicians, such as Minnesota's future

US Vice President Hubert Humphrey, joined the fray, emphasizing the importance of these projects to their region. While the MURA was pressing on the Fixed Field Alternating Gradient machine and the community was not impressed with the value of this machine, the midwest was in for a pleasant outcome.

The regional scientific war which the *New York Times* called the "toughest ever fought for a federally sponsored installation," was transformed when the campaign for a TNL was merged with alliance of politicians and MURA, and when a more national Universities Research Association (URA) was formed with 34 leading University members. The "war" was won by the URA and a new laboratory, National Accelerator Laboratory (NAL), now named Fermi National Accelerator laboratory, was approved in Weston, Illinois.

The Fermi National Laboratory insiders firmly believe that the politics of "Open Housing" was very instrumental in the final choice. The NAACP (National American Association of Colored Persons) and Martin Luther King were agitating against racial discrimination prevalent in renting and sale of housing to colored people and wanted new laws to ban such discrimination. These insiders believed that Johnson got the support of a Republican Party Illinois Senator Everett Dirkson, who vehemently opposed the open housing laws, only after Weston, IL, was chosen as the site for the would-be NAL. This might be true given the pithy policy statement

Fig. 7.14 (Fermi) National Accelerator Lab Main Ring (Courtesy: Fermi National Accelerator Laboratory)

Fig. 7.15 Main ring dipole
magnet during the
deconstruction stage.
(Courtesy: Fermi National
Accelerator Laboratory)

by the would-be Director of the lab, Robert Wilson: *"...we have observed the destiny of our Laboratory to be linked to the long history of neglect of the problems of minority groups. We intend that the formation of the Laboratory shall be a positive force in the progress toward open housing in the vicinity of the Laboratory site."* Robert became the first Director of the NAL, because of the peevishness of Edward Lofgren at BNL, who was offered the job, at having to build the "Berkeley Machine".

Robert Wilson was a physicist who left his deep footprints in physics, though he had many false steps in the first few years. A very talented man, he invented experimental devices, such as sliding vacuum seal to enable inserting objects into a vacuum chamber without breaking vacuum and, at the same time, did complex theoretical calculations on cyclotron orbits. (Yet, strangely, the man was fired from Berkeley Radiation lab twice for being clumsy and careless.) His work on proton radiation therapy still provides the basis for treatments in major centers. The colorful man from Frontier, Wyoming, was a celebrated sculptor, a passionate advocate for human and civil rights, an individualist, and yet one who was a team builder. Wilson had led a group in the Manhattan project in Los Alamos, then gone on to build the world-class particle physics program in Cornell University, and led development of accelerators there, particularly electron accelerators. His physics vision and justification on the exact need for synchrotron accelerators, his technical knowledge of subtleties in such accelerators, and appreciation of aesthetic nature of experiments were inspiring. He was also a consummate leader who could motivate, influence, fraternize, direct, organize, and execute. So this "frontier man" was the natural choice to lead the construction of the present generation of synchrotrons. The later Director and Nobel Laureate Leon Lederman would say this of him, "*His spirit invades every corner of this great laboratory. He speaks to us through the surfaces of precast concrete, through the prairie restoration, through the style of openness, through the flags that grace the entry.*"

In the late 1960s and early 1970s, this combination of talent in leadership was critically necessary, because in the USA, leading physicists were convinced that larger and larger machines with complex components and systems were becoming essential to advance the frontiers of physics. While most theoretical physicists understood and applauded such projects, not all experimental physicists were sure. Many of them were uncomfortable with the impossibility of knowing everything about the "apparatus" that they would be using as they had done in the past and then there was also the fear that such investments would drain funds away from smaller experiments on "basic" aspects. To be candid, this tension between "big science" and University scale science persists even today and much bargaining is done before the community comes together on such decisions.

In November 1967, the National Accelerator Laboratory came into existence. With Wilson in charge, the machine design changed considerably with significant cost reductions. He led his team in coming up with an approach that was used in small machines, and endowed it with holistic, simple, and inexpensive methods. The team also worked at a high speed, consistent with the energetic temperament that Wilson had acquired during his association with Lawrence. The groundbreaking for the first injector Linac took place in September 1968 and the machine (the Main Ring) achieved 200 GeV in March 1972, and 400 GeV in December 1972, an amazing achievement, given the complexity of this machine. In May 1974, the NAL was rededicated and renamed Fermi National Accelerator Laboratory (FNAL).

The proton beam, in this cascade of accelerators, (Figs. 7.14 and 7.15) starts its life as a negative hydrogen ion (H^-) beam from an ion source, and a Cockroft Walton generator accelerates it to 750 keV and a 400 ft (130 m) long Alvarez type linear accelerator (see Chap. 8) with six accelerating sections accelerates it to 200 MeV. The beam then passes through a carbon foil, where the H^- ion sheds two electrons to become the proton. A fast cycling, 1,600 ft (about 480 m) circumference booster synchrotron accelerates the beam to 7 GeV and fills the Main Ring synchrotron with 12 pulses of protons. The proton beam is then accelerated in the Main Ring to the full energy. Wilson adopted many ideas of Sands and the WAG group. The two main innovations by Wilson that allowed the machine to be cheaper, simpler, and yet more powerful were – (1) The magnets became separated by function – a set of dipole magnets did the bending and the D and F quadrupole magnets (see above) provided the alternating gradient – strong-focusing, (2) the idea of a cascade of injectors including the booster ring. These two aspects have come to stay in modern circular accelerators.

The 4 mile (6.4 km) circumference main ring was the first synchrotron to seem straight when one peered along the beam line. In such a design, the machine might as well be linear, since the beam tube diameter, the local vacuum, and electrical needs, etc., do not depend upon the machine radius. The Main Ring consisted of 774 bending and 240 quadrupole magnets, each magnet up to 20 ft long and weighing up to 12.5 tons. The last of these was installed in April 1971. It is a credit to all the accelerator physicists, past and active then, that such a large and state-of-the-art machine could be built without any serious glitches and problems. On March 1, 1972, the protons were filled from the booster at 11 A.M. At 11:30 A.M., the beam had crossed the transition gamma and at 1.03 P.M., Stan Towzer declared "That one went all the way out." The proton beam had reached 200 GeV, traveling 70,000 times around the ring in 1.6 s. The FNAL history website (Fermilab History Project, Accelerator, and Main Ring) describes the celebration,

"In the Control Room on the next pulse, someone in the hushed crowd said, "There it is!" and a rousing cheer filled the room at 1:08 pm!

On a desk in the lobby sat a carton with a white handwritten label reading, "For Ned Goldwasser. . .for 200 GeV celebration. . .from Al W. . . It's the correct brand. Tradition calls for 40 persons per bottle at lower energy machines...." Edwin L. Goldwasser, Deputy Director of the Laboratory, ordered the carton opened. The gift of chianti wine came a few days before from Al Wattenburg, professor of physics at the University of Illinois, who was one of the small group present at the first self-sustaining nuclear chain reaction achieved in 1942 by the team headed by Enrico Fermi, when a bottle of the same brand of chianti was passed among that group of pioneer nuclear scientists. Now, thirty years later, another group of jubilant scientists shared a major achievement in particle physics. Dr. Wilson and Dr. Goldwasser passed through the crowd filling paper cups, shaking hands, accepting and extending congratulations at every turn. Later, champagne that had waited for many months in the cafeteria cooler, was served in plastic goblets labeled "200 GeV"!"

As with human rights issues, Fermilab also became a good steward of the land they had acquired for the laboratory. The native prairie land was restored, including bringing in a herd of buffalos (bisons) that used to roam in large numbers in the past. Green spaces were created and old farmhouses converted to idyllic guesthouses for visiting scientists. (This practice has now become the norm for accelerator laboratories around the world.) After serving for about 25 years, the Main Ring gave way to the Main Injector for a larger machine, again a first in the accelerator world.

Not too long after the success of the Main Ring, CERN also celebrated with its own SPS (Super Proton Synchrotron – not a superconducting machine), a comparable synchrotron which, like the main ring, is giving excellent service as an injector for a larger machine. The 6.9 km (about 3.9 mile) circumference machine exceeded its goals and achieved 400 GeV and 10^{13} protons in December 1976, half of which were extracted to the outside. But the Fermilab main ring had already exceeded the energy by running at 500 GeV about 6 months or so earlier.

The change from a beam bound within a magnet to a ring of beam tube has liberated the size of the ring. As long as resources and space can be found, there is now full confidence that these rings of the earth can be built. Indeed, hundreds of synchrotrons varying in diameter from several meters to several tens of kilometers in circumference are in operation. They serve varying purposes, some providing high-energy X-rays, some providing medical isotopes, and many in the service of high-energy physics. Like the rings of Saturn, they may break away in the far future, but while they are here, they add an aesthetic sense to human purpose.

Chapter 8
The Next Generation: Supersynchrotrons

The advent of the alternating gradient synchrotrons and linear accelerators established the potential of limitless energies with an intense beam in colliders. Experience with the strong focusing machines brought accelerators to such a level of maturity that there is now a very clear prescription for physics and engineering design, construction, and operation. Accelerator physics, with respect to beam dynamics, diagnostics, and machine utilization, and the technology of magnets, had been more or less established. The use of quadrupole magnets for focusing further cemented the design approach for high-energy accelerators. What remained was to push the required technologies to achieve energies once considered pipe dreams. One big push came from a different field of physics. With this enabling technology, the accelerators are quite feasible and costs are not unreasonable at least for energies below tens of TeV (10^{12} eV).

It's a Bird, It's a Plane, It's Superconductors

The feasibility to these high energies is made possible by one important invention – practical superconducting devices. Clearly at these very high energies, the strength of the magnetic field needs to increase proportionally, high intensities require large bore magnets, and the ring needs to be very large. The power and energy consumed by conventional magnets increases with the square of the magnetic field and square of the bore diameter. So 1 TeV particle energy would require 100 times the power of a 100 GeV magnet or the synchrotron radius would have to be increased 10 times with 10 times more number of magnets. For linear accelerators, either the beam line length would have to be increased 10 times or the accelerating cavities would have to be 10 times more powerful.

The ability to build to TeV energies is now made possible with superconducting magnets to provide high magnetic fields of high field quality and adequate bore size to accommodate the beam and superconducting radiofrequency (RF) cavities to provide large accelerating voltages without accompanying large losses. The new era of superconducting accelerators is now here to stay, so much so that even earlier concepts of accelerators such as cyclotrons, being used in medical applications, have adopted the superconducting technology. However, the superconducting accelerators utilize the same basic principles of synchrotron as previously described accelerators. The use of superconducting technology brings down significantly the power and the power supply and changes cooling requirements for high field magnets and enables high accelerating voltages for RF cavities. Superconductors can carry currents several orders of magnitude higher than normal conductors, with near-zero power consumption, so that very powerful magnets can be built.

Superconductivity and Superconducting Devices

All conducting materials work from the principle that the solid metal has atoms fixed in a so-called lattice structure which can be physically imagined as an arrangement of fixed atoms, say at the edges of a cube. In the environment of this lattice of highly conducting metals, the electrons in the atoms can become detached from the atoms and can move freely in the neighborhood while maintaining overall charge neutrality. This free electron "gas" is a conducting medium and starts carrying an electric current when a voltage is applied across the conductor and drifts towards the voltage terminal due to the electric field. But in doing so, the electrons collide with lattice atoms which are constantly vibrating in their positions due to thermal disturbances. These collisions take energy away from the electrons and therefore the river of electrons feels the friction of having to move through this "rough" surface of jostling atoms. This friction is the electrical resistance of the conductor and is an inherent property of the metal. To overcome the resistance and sustain the electric current, a part of the voltage that was used to start the current has to be maintained. Just as an object heats up when it is moved against friction, the lattice atoms heat up due to the energy received from the colliding electrons and vibrate even more. This "Ohmic" heat has to be removed by efficiently cooling the conductor or the resistance would increase further and if the cooling is insufficient, the conductor would melt and break. The insulating material that separates conductors of a coil would also heat up and breakdown, causing short circuits. Cooling channels increase coil size and result in inefficient and unwieldy magnets. For large magnets, a large current and a large amount of conductor would be required. This would increase the power requirement, and the amount and size of equipment to supply this power and to cool the magnet might become unacceptably expensive or infeasible. (In some cases, the requirements can be lowered by providing special coolants such as liquid nitrogen to lower operating temperature and therefore the resistance of the coil.) The superconducting technology overcomes this limitation.

In 1911, Kamerlingh Onnes discovered that when cooled, mercury, which is only a moderately good conductor of electricity, suddenly dropped to zero resistivity at the critical temperature of 4.2 K (0 K corresponds to $-273°C$). He called this new state "superconductivity" and he and others received Nobel prizes on the experimental and theoretical discoveries for this phenomenon. This fantastic new state of matter comes about because in some metals, the free electrons form what

Fig. 8.1 The critical field at
different temperatures for soft
(type I) superconductors

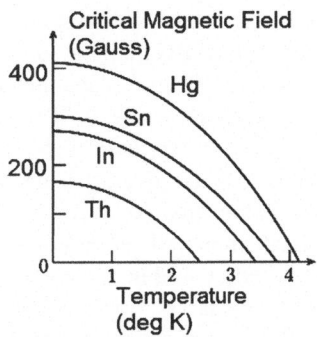

are called Cooper pairs and "dance" through the metal lattice in such a way that
when one loses energy by interaction with the lattice, the other gains energy from
the lattice so that the overall energy loss is zero. However, this superconducting
state remained a scientific curiosity because the metals transitioned to normal state
even at very low magnetic fields of a few hundred gauss (see Fig. 8.1).

In early 1960, there was a new development where hard metals such as niobium
were found to exhibit a more robust superconductivity. These have so-called mixed
states with some normal and some superconducting regions so that the whole metal
did not go normal at some critical magnetic field. This situation was further
improved with alloys such as niobium–titanium and niobium–tin alloy or composite
superconductors. With these very significant inventions, it is now possible to
increase the amount of current to very large values (by many factors to thousands
of times more depending upon the magnetic field) and yet not incur any resistive
heating or loss of power. Indeed, in MRI magnets used in medical applications, the
electric current, once initiated in the magnet coil, is sustained for decades if the
electrical joints that close the loops are also made superconducting. These
superconductors can work up to fields of 250,000 gauss (25 T). The alloy and
composite superconductors (mixed superconductors) have to be cooled to 20°K
(−253°C) or below for achieving superconducting state. The most used
superconductors are niobium–titanium (magnets for lower field accelerator and
fusion, MRI magnets, superconducting motors, etc., which require liquid helium
or lower temperatures) and niobium–tin (high field magnets and devices that can be
operated at temperatures below 10°K that is −263°C). High temperature
superconductors, some even at room temperatures, have been discovered more
recently. Although these are used in some applications, the technology of material
and fabrication is not yet mature for large-scale use in accelerator magnet coils.

Cool Currents of Superconductors

Alloying or mixing and sintering niobium with other metals such as titanium, tin,
aluminum, etc., makes superconductors which can then be drawn into filaments.
Though these superconductors have zero resistivity when cooled down to very low

temperatures below critical temperatures, there exists a maximum critical current for a given magnetic field and temperature above which the superconductor would transition to a normal state. This critical current decreases as the magnetic field or temperature is raised. If the superconductor goes normal because current or temperature or magnetic field is too high, the conductor would jump to the "normal" state with a high electrical resistance and a large Ohmic heating would result locally in that normal zone. If the circuit remains closed, the current will continue to flow till the magnetic energy in the coil is dissipated. This, and any voltage that is being applied, would heat the coil locally substantially, because the coil current densities are very large. However, well before the current has decayed, the superconductor would burn up very quickly. Such "quenching" of superconductor can also be caused by small instabilities in the cooling environment or sharp jumps in magnetic fields. The superconductor quench can have disastrous results. Besides the burn up, large "inductive" voltages would arise across the coils due to sharp change in the current due to the resistance or break in the circuit due to the burnout. Such inductive voltages could cause further breakdown in the coils (in the insulating materials) and might even damage other magnets in the same string.

To improve the superconducting stability and avoid catastrophic quench, the superconductor filaments of the niobium–titanium or niobium–tin are embedded in highly conducting copper to form a matrix which stabilizes the superconductor. So if and when a quench occurs in the superconductor and it becomes highly resistive, the current jumps into the lower resistivity stabilizing copper which provides a bypass for the current. However, typically, the magnetic energy is too large and one cannot wait till the current has decayed to zero, since even the copper matrix might burn up. Therefore, once a quench is detected from the smaller inductive voltage, the circuit is disconnected and the current is bypassed into another circuit with "dump" resistors. Simultaneously, the coil is forced, with additional heaters to quench all over, so that the stored energy is distributed more uniformly. Figure 8.2 shows a typical cross section of a superconducting strand, which is used in accelerator magnets. Such strands are also used in thermonuclear fusion magnets. While these strands have diameter of the order of a mm, the MRI superconductors are much larger. The superconducting filaments inside the strand are made as fine as possible (a few microns), because while the current is ramped up or reduced,

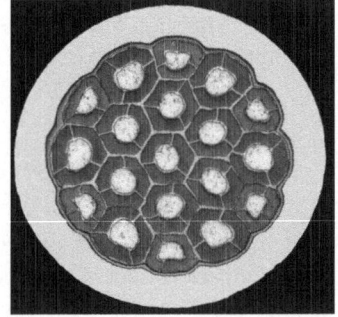

Fig. 8.2 Cross section of a niobium–tin strand after heat treatment. The small filaments of niobium which has reacted with the central pool of tin to form the alloy can be seen on closer inspection. The sheath of stabilization copper can also be seen. (Courtesy: Philips Healthcare, USA)

so-called shielding currents would be induced, affecting the stability and increasing magnetic field error (field quality). The larger the size of the filament, the larger would be these currents. Such "persistence" currents also prevent the equal sharing of the current between filaments so that some filaments can end up with too much current and quench. For accelerator magnets, several strands are braided into ribbon shapes. The specific braiding pattern is controlled or specific coatings are applied on the strands, because while ramping the coil current, additional currents between adjacent strands might be induced, destabilizing the superconductor and affecting the equal distribution of strand currents. Even with the stabilizing matrix and quench detection system to protect the conductors, the actual reliable operation of the magnets requires mechanically and thermally shielded configuration. At the low operating temperatures of superconducting magnets, the heat capacity of the materials (the capacity to receive heat with only small rise in temperature) is extremely small and any frictional or external thermal heat has to be minimized.

Basic Design of an Accelerator Magnet

The strong focusing AGS and the PS had the C shaped magnets, in which a coil is wrapped around iron and the field is created at the gap between the poles. The C shaped magnet with alternating orientation (facing in or out) gave the alternating gradient with appropriate shaping of the poles. Once the quadrupole magnet became the focusing elements, another design, the "picture frame" magnet became common (Fig. 8.3).

The current flows in one direction in the left conductors and returns through the right conductors and the ends of the magnets have the u bends of the conductor turns. The frame is made of iron and in this way it forms a mini H type magnet. The

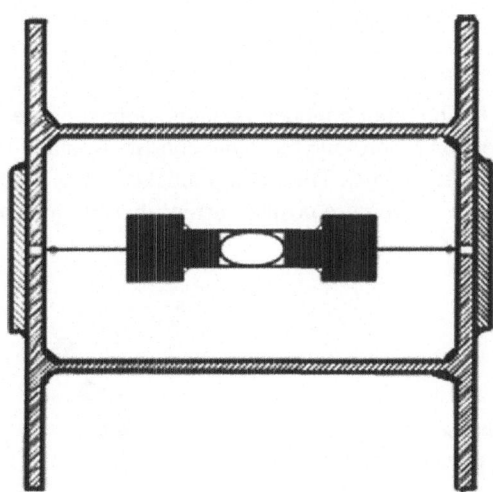

Fig. 8.3 Fermi Main Ring (Injector) dipole bending magnet ("*Starting Fermilab*", Robert R. Wilson, Fermilab Golden Books, courtesy, Fermi National Accelerator Laboratory)

iron yoke is useful only for fields of up to 1.5 T, since it saturates and then the fields are provided as if there is no iron pole. When high fields are required, the conductor should be as near to the field region as possible, and therefore should be as small as possible, for a given beam tube space. In a rectangular picture frame, the corner conductors are farther and therefore less efficient and also this type of coil gives rise to a nonuniform magnetic field. It is preferable to put the conductors in a circle around the beam.

For a pure dipole magnetic field, the field in the region needs to be uniform and in a single (vertical) direction and for the quadrupole magnet, the field has to increase exactly in proportion with radial distance from the center. Theoretically, this type of pure field can be achieved with the so-called Cosine coils–coils in which the current in a conductor (or the current multiplied by number of turns) is proportional to the cosine of the angle of location of that conductor for the dipole and cosine of twice the angle of location for the quadrupole magnet.

The field (in Gauss) at the point due to the conductor C is given by $B = I/(5R)$, where R is the radius of the coil (cm) and I is the current in the conductor (Amps) and is perpendicular to the line connecting the point to the conductor (R in Fig. 8.4). With y direction as the vertical, the y component of the field

$$B_y = I \cos(\theta)/(5R). \tag{8.1}$$

If current in the conductor is also arranged to vary with the angular location of the conductor such that $I = I_0 \cos(\theta)$, where I_0 is the current in the conductor at the median plane ($\theta = 0$), then

$$B_y = I_0 \cos^2(\theta)/(5R). \tag{8.2}$$

When we integrate the field over all the right side conductors located around a semicircle ($-\pi/2$ to $\pi/2$), we get

$$(B_y)_{\text{right}} = \pi N I_0/(5R). \tag{8.3}$$

The set of conductors on the opposite side (the sign of current as well distance is opposite) provide the same contribution and therefore, the total dipole field $(B_y)_{\text{total}} = 2\pi N I_0/(5R)$. The x component of the magnetic field, $B_x = I \sin(\theta)/(5R)$, is exactly zero in the pure cosine current distribution, because the contribution from the left

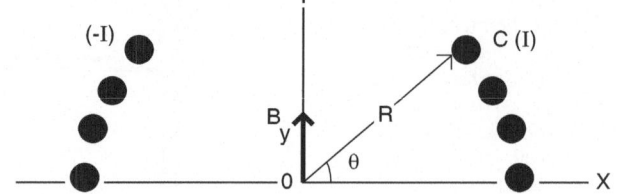

Fig. 8.4 Magnetic field at the point O due to the conductor with a current I at radius R

side cancels that from the right side. Therefore, the field is purely vertical, bending the particles in the horizontal plane. A similar calculation can be carried out for the quadrupole, which would have a current variation of $I = I_0\cos(2\theta)$ with four poles. This gives a pure gradient field which is zero at the center and increases linearly with radial distance.

However, in actual magnets, this is hard to achieve since the coil is built with discrete set of conductors and the current (actually ampere turns) around the circumference cannot be made to vary smoothly as a cosine of the angle. The conductor also has a substantial radial width (one to several cms) which may be a significant fraction of the beam pipe radius. Therefore, the optimum locations of the conductors are calculated using a code for a given size of the conductor to obtain as pure a field as possible. A few wedges (spaces with no current) are also placed in the coil to enable more freedom to locate the conductors. Though, this may seem to be an inefficient use of the conductor space, without these wedges, it is, in general, not possible to obtain pure dipole or quadrupole fields. The coil turns and even the distribution of strands of the ribbon have to be accurate to micron level to obtain high field quality (purity of vertical or gradient fields). Since superconducting magnets and the RF sources are feats of engineering, not unlike space vehicles requiring state-of-the-art material technology and engineering, extraordinary precision and quality control to achieve over 99.99999% reliability, it is worth describing the fabrication process (Fig. 8.5).

High field coils are made with niobium–tin strands, which become superconducting only after heating (curing) to 1,200°C for 30 min. First these strands are woven in a specific pattern into a thin ribbon, which are then insulated

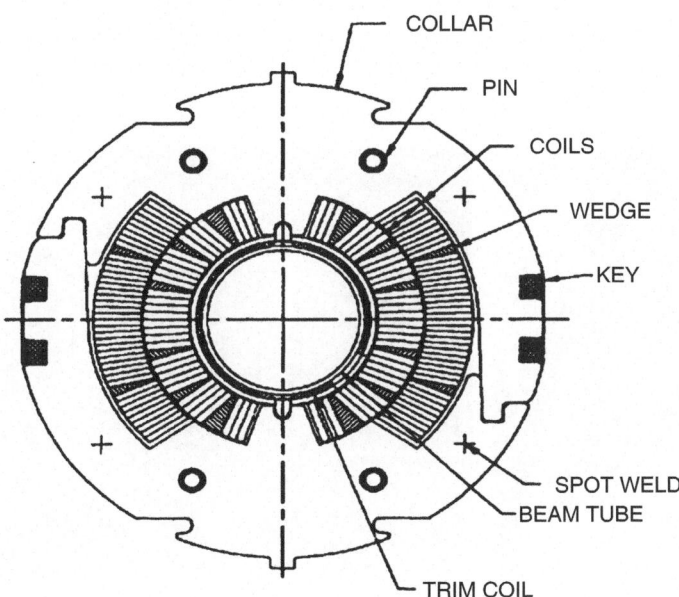

Fig. 8.5 Cross section of an SSC dipole coil. SSCL Site Specific Conceptual Report, July 1990

with radiation resistant film insulation and then with epoxy impregnated fiberglass tapes. Since the strands become extremely brittle, the insulated ribbons are wound with insulated wedges spacers in a winding tool and with end spacers, then held in a special tool which puts the coil assembly under pressure and cures (with the setting of the epoxy) at high temperatures. The tool dimensions and the process of winding and curing are controlled to a great precision, in order to obtain the coil locations within tens of microns of the code calculations. The coil is made in the form of race-track-like saddle coils with long straight sections and short ends. Figure 8.6 shows a fabricated coil and the curing tooling. Lower field niobium–titanium coils are made with ribbons of strands and wound into coils and used directly.

Typically, a dipole magnet is made by clamping a top and a bottom coil between two metal collars. The collars are made from a number of laminates of stainless steel (to avoid the so-called eddy currents which might be induced when the magnetic field changes during the current ramp up or down). The clamped coil assembly is then placed between two halves of iron laminate yokes. The iron yoke helps to obtain a higher magnetic field and to prevent the leakage of magnetic fields outside the magnet assembly. The whole assembly is then placed between two aluminum or stainless steel semishells under compression and welded together. Welding shrinkages and cool down to low temperatures cause the assembly be held in compression. The art of making the superconducting coil is intricate requiring precise engineering modeling of both the design and the process. The fabrication process has to be carried out with meticulously chosen materials which must withstand high currents and magnetic fields, high potentials in case of magnet trips, high mechanical pressures, and a large range of temperatures from 1,200°C down to −270°C. The splice joints between coils, for example, between the upper and lower coils are designed and developed with great care to withstand the large mechanical forces exerted on them by the complex magnetic field in that region,

 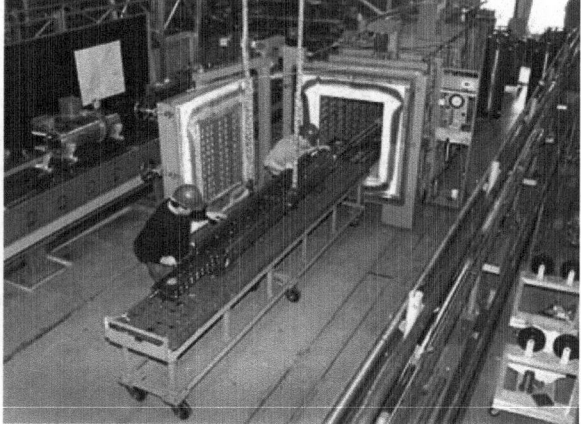

Fig. 8.6 (*Left*) Saddle coil of a dipole magnet (*Right*) A niobium–tin coil being loaded into a curing (reaction) oven. (Science@Berkeley Lab, Oct., 23, 2007, Courtesy: Lawrence Berkeley National Laboratory)

while the electrical resistance of the joint is minimized. (This is one of the most sensitive and least reliable parts of the magnet which can, as it happened in the LHC machine, cause major shutdowns.) All these parts of the magnet have to withstand large doses of nuclear radiation over the operating life time. Special large tools (with intimidating sizes) and jigs are fabricated to achieve the design parameters in fabrication.

The iron yoke laminates have holes for the circulation of liquid helium which is needed to cool down the coils to the operating temperature of 4°K or lower and to remove any small heat that may be generated by mechanical movements or somewhat resistive components like joints between the turns. During operation, typically the magnets receive minimal amount of external heat load (for example, the dipole magnet of the supercollider would have received about 2 watts per magnet due to synchrotron radiation – radiation emitted by proton in moving in a circular orbit and any additional heat caused by so-called hysteresis losses when current is ramped in the magnet). Since material (even copper) thermal conductivity becomes very small when cooled to very low temperatures, one cannot rely on conduction through walls of the coil to cool the coil, and therefore the helium has to flow right near the coils. The allowed temperature rise in the cooling helium is extremely small, of the order of 0.01°C since there is very little temperature margin for the superconductor – meaning the superconductor would go normal with even a small rise in temperature. Therefore, even the few watts of heat load can be onerous and the cooling has to be exquisitely designed and confirmed.

The whole cold mass is enclosed in a cryostat which is like a thermos flask, providing a vibration resistant, stable, and strong support while preventing any heat leakage from the external environment. The cold mass support consists of multiple steps of composite thermally insulating materials arranged in such a way that the path for heat leakage from outside is very long. The cold mass is draped in reflecting foil to prevent receiving any heat radiation. There are two shells in the cryostat, an outside iron shell at room temperature (steel confines all magnetic field so that external instruments may not be affected) and an inner aluminum shell which is maintained at liquid nitrogen temperature (77°K or −196°C) so that the surface "seen" by the cold mass is at low temperature and heat radiation is reduced. The use of such a "shield" reduces the requirement for the more expensive liquid helium (Fig. 8.7).

As an example, the Superconducting Supercollider, which was not built, had dipole magnets of advanced design that produced 6.6 T (66,000 gauss) at a current of about 6,500 A in a bore of 5 cm diameter and were 15 m long with an outer diameter of 66 cm. There would have been over 8,000 such magnets in the collider. The quadrupole magnets produced a magnetic field gradient of over 200 T per meter radius and had a length of 5.2 m. All these numbers are spectacularly large, requiring a large research and development effort. The Large Hadron Collider Magnets (see later chapter) have even larger fields with larger bore and are two magnets in one. The superconducting accelerator magnets are a marvel of techno-logical achievement combining precision with cutting-edge materials based on advanced physics discoveries. The design and development of materials, tools

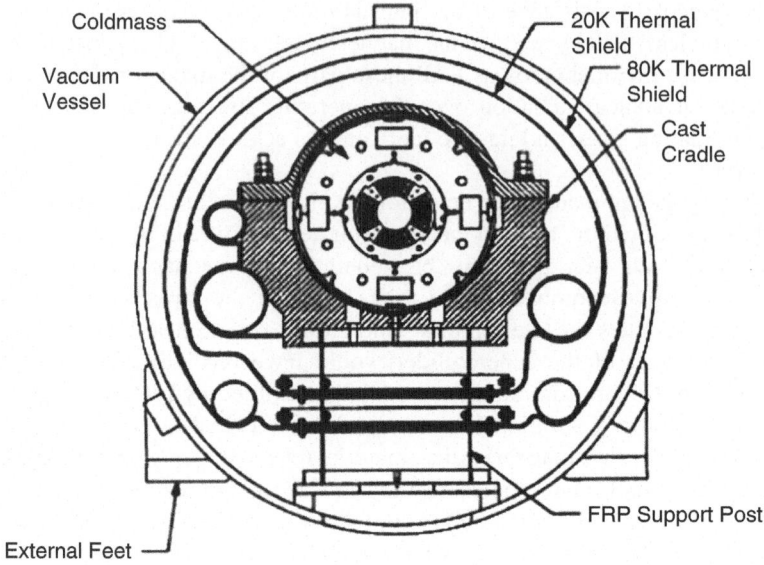

Fig. 8.7 Cross section of a superconducting quadrupole magnet, SSC Site Specific Conceptual Design Report, July 1990

(tooling), and processes for the manufacture of such magnets is an amazing achievement of the 1980s and 1990s. Over the years, such experience in superconducting design and fabrication had been gained in the Tevatron, HERA, NUKLOTRON, and RHIC accelerators. This experience and that gained in the Supercollider project provided considerable learning experience for pushing the limits of the magnet technology further such that very advanced magnets such as in LHC are real today.

In addition to the dipole and quadrupole magnets, correction magnets, which correct field errors in the dipole and quadrupole magnets, are also made from superconducting wires. Since the correction fields are usually small but have complex winding patterns, the choice of superconductor is more to avoid losses and make the magnets compact. The technology of the correction magnets has evolved for achieving this purpose and often the coils are patterns placed on insulation and rolled up to get the desired field configuration.

Superconducting Radiofrequency Accelerator Cavities

The Radiofrequency (RF) cavity is where the particles increase their energy by being accelerated by an applied electric voltage. The RF voltage, as stated before, is synchronous with particles to continuously accelerate the particle bunches. The bunches (buckets) are separated in space such that when one bunch arrives, the other leaves. This way, each bunch sees an accelerating voltage as it enters and a repelling voltage as it leaves. In addition to accelerating the particles, the accelerator cavity must also replace the energy lost by particles through synchrotron radiation which the particles emit when they do not move along a straight line. (Therefore, this is generally an issue for synchrotrons and not for linear accelerators.) The RF cavity design has come a long way since the days of Wideroe, aided by extraordinary developments in the application of superconducting materials. These developments, while important for circular accelerators, are the key to linear accelerators of the future since so many of them would be required for them.

The challenge of the RF cavity is to generate high accelerator voltages so that particles can be accelerated without going through too many turns. The demand on the voltage is particularly severe for linear accelerators where multiple cavities are used along the accelerator length, which can be made shorter, reducing size and cost, if the accelerating voltage per unit length of the cavity can be increased. When these voltages are large, considerable energy loss occurs in the RF structure and electrical breakdowns are a serious issue. The total power that must be fed into the cavity is the product of the cavity voltage and the beam current (number of particles crossing the cavity per second × electric charge). One way to limit the energy losses is to make the cavity structure that confines the strong radiofrequency electric fields, out of superconducting materials, reducing power requirement to near zero. Superconducting cavities have zero D.C. resistance and therefore can achieve high "Quality Factor," ratio of inductive to resistive voltage.

When the development of superconducting cavities began, initial results were limited to a few MV/m electric fields. At these fields, considering the cost of cryostat, etc., these would not have been competitive. But, recent developments in design, material, and clean room fabrication, including surface deposition techniques, have pushed the gradient to near 40 MV/m, a stupendous number. Central to this achievement are the material processing techniques, viz., chemical cleaning of the surface of the cavities, removal of any specs and sharp burrs which would locally increase the electric fields, causing field emission and electrical breakdown, annealing of the superconducting niobium material at 1,400°C to increase its thermal conductivity to maintain superconducting stability, and rinsing methods to remove surface contaminants, etc. Since such cavities operate at very low temperatures, these too need a cryostat to minimize external heat. The simplification of cryostat design has been an important factor in this feasibility. While the normal cavities operate at a frequency of the order of 10 GHz, superconducting cavities operate at a lower frequency (around 1 GHz), making the sources simpler and mechanical tolerances easier to achieve. The lower frequency also allows long particle bunches (Fig. 8.8).

While niobium–titanium and niobium–tin are better superconductors, the maximum gradients these could sustain are limited partly due to their granular structure and grain-boundary effects and because these materials are affected more by high frequency fields. Superconductors are nondissipative in DC conditions, but when alternating fields are applied, the hysteresis involved in their persistent currents causes dissipation, much like in iron and this dissipates heat, causing

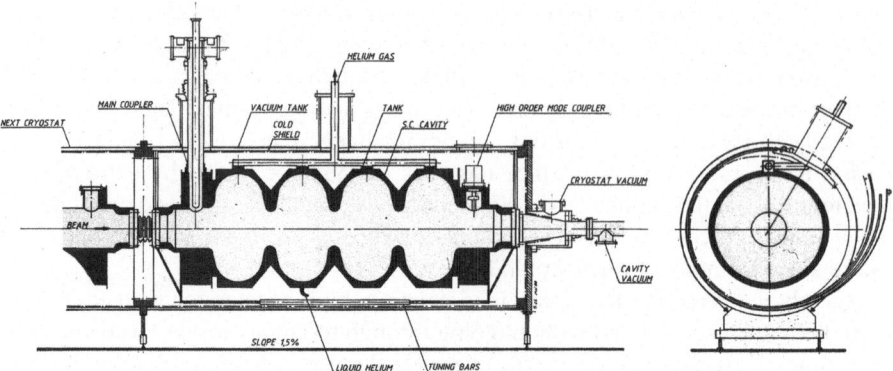

Fig. 8.8 Example of a superconducting cavity inside a cryostat for the Large Electron Positron Collider II (Courtesy: CERN, European Organization for Nuclear Research, Image Reference: CERN-DI-9107011)

superconducting instability. The energy dissipation is given by the equivalent AC resistivity,

$$R_{AC} = (k\omega^2/T)\exp(-1.76T_c/T),$$

where T_c is the critical temperature (temperature above which the superconductor would turn normal), T is the operating temperature, and ω is the RF frequency. Due to the exponential dependence on the temperature and the dependence of the proportionality constant k on temperature, the AC resistance drops by nearly two orders of magnitude, if the cavity can be operated at supercooled liquid He (LHe) temperature of 2 K rather than the 4.2 K. The maximum sustainable RF field is also limited by the magnetic field associated with it and even earth's magnetic field affects the cavity and shielding is required to achieve high RF fields. Niobium is the usual choice for the RF cavities, either in the form of thin sheet or sputtered onto copper surface. Accelerators, such as the Cornell synchrotron and DESY, have used sheets while the Large Electron Positron Collider in CERN (Fig. 8.8) used sputtering technique. The cavity is designed carefully using specialized codes to get most optimum contours of the surfaces while giving the correct resonant structure avoiding undesirable modes of electric field oscillation. The design and fabrication is both a science and an art and the gleaming product is usually a testament to the RF cavity scientist-sculptor (Fig. 8.9).

Supporting Cast-Facilities and Systems Engineeing

As stated earlier, the superconducting magnets and RF cavities are required to be cooled to very low temperatures and typically this is accomplished by circulating liquid helium through or around the devices. Therefore, the accelerator complex

Fig. 8.9 The ICHIRO5 superconducting RF cavity, designed by KEK, Japan, which achieved 33 MV/m electric field at the Jefferson Laboratory (Courtesy: Thomas Jefferson National Accelerator Laboratory, VA, USA)

would require large liquid helium (and liquid nitrogen – for cooling shields and other components) plants and associated distribution systems. The accelerator(s) are typically housed in underground tunnels and so these areas would have to be serviced with elevators and people and material movers. A complex array of electrical supplies, electrical busses, sensors, and sensor arrays are required. A complex control, operation, and monitoring system is needed to operate the accelerator efficiently and safely. Finally, an enormous amount (hundreds to thousands of gegabytes) of experimental and monitoring data would be gathered during all phases of the operation and sophisticated data management, analysis, and visualization system with highly advanced information technology support (see Chap. 11). Each of these support systems is large by any standards and require considerable planning, design, fabrication, installation, and maintenance usually with impeccable reliability.

One of the key concepts developed through years of managing such large project is the Systems Engineering and Quality Assurance approach. This is well described elsewhere in management books. But suffice it to say, that like the space programs requiring a reliability of 99.99999% or higher, the accelerator programs are an example of human achievement in precision and reliability. This reliability is achieved by ensuring that device and component requirements are threshed out to meet the operation, the design is reviewed at every stage and judged with regard to satisfying requirements. Risks involved in not meeting the requirement or in failure are assessed; operational safety, feasibility of timely and on-budget fabrication are assured; and a design description that ensures ease, quality, and accuracy in fabrication. Once the design is finalized, the Systems Engineering approach freezes the design and creates a configuration management system consisting of all documents pertaining to the design requirements and the design description. Any change to the configuration would then have to be reviewed by a group of scientists for approval or denial. The fabrication and assembly of the devices follows this configuration and the process of manufacturing would incorporate all quality assurance methods to ensure the required reliability and would also document all processes and in-process testing results. The latter assists in tracing an offending device or component if there is a failure. Same approach is adopted in commissioning and operation so that all steps are deliberated, reviewed, planned, and controlled for conformance to the plan and safe and efficient performance.

Tera on Terra: Tevatron to Large Hadron Collider

The first of the supersynchrotrons was the Tevatron, the second highest in energy in the world, which had all superconducting magnets, but a normal RF cavity. HERA in DESY is the first machine with both superconducting magnets and superconducting RF cavity.

Enrico Fermi Would Be Proud

In 1972, the Fermi National Laboratory (FNAL) achieved the first FODO cell based synchrotron, to the great delight of the accelerator physicists around the world. Following the Main Ring, it had nowhere to go because in 1976, the decision was taken by the High Energy Physics Panel (HEPAP) and Department of Energy, in US government, to build the next machine (a collider named ISABELLE) at the Brookhaven National Laboratory (BNL). Fermilab Director Robert Wilson resigned in protest. In 1978, Fermilab was merely "in drift." But in 1978, the ISABELLE project ran into trouble because the superconducting magnets had serious technical problems. While BNL scrapped this project and proposed another, Nobel Prize winner and the new director of FNAL, Leon Lederman, grabbed the opportunity and on November 15, 1978, secured the funding for doubling the energy of the Fermi Main Ring, using superconducting magnets. The new machine, 3.9 miles (about 6.1 km) in circumference has been continuously upgraded ever since such that it has now achieved the 1 TeV energy in each of the proton, antiproton beams (circulating in the same beam tube), deserving its new name Tevatron. In early 1984, the Tevatron achieved the phenomenal 800 GeV energy and in 1986, 900 GeV. In late 1986, proton antiproton collisions were obtained at 1.8 TeV total energy. After several new results in particle, FNAL achieved its claim to fame by detecting the long predicted Top Quark (Fig. 8.10).

The program of developing superconducting magnets and an accelerator system to accelerate particles at a rate of 50 GeV per second had been started in 1972. The Tevatron has a beam tube with 7 cm diameter and a magnet bore of nearly 8 cm diameter. There are 774 bending magnets, each 6.4 m long with a field of 4.2 T, and

Fig. 8.10 Aerial picture showing the Tevatron at Batavia, Illinois, USA (Illustrated with *red* line). The larger ring is the Tevatron and the smaller one the Main Injector (Courtesy: Fermi National Accelerator Laboratory, IL, USA)

216 quadrupole magnets around the ring, operating at about 4 kA. The total amount of superconducting wire used in that magnet was 42,500 miles, an astounding production achievement at that time. Twenty-four local refrigerating units were employed initially and were upgraded later. The cold boxes are over 13 m tall, with four 4,000 horsepower compressors pumping liquid helium at a rate of over 250 L/min and correspondingly large liquid helium and nitrogen storage systems. The other achievement of this facility was the production and storage of the antiproton beam. Any one, who has watched SciFi episodes like Star Trek, knows that this is the stuff of explosions, engine drives, and so on. Since any contact with matter containing proton will cause annihilation of the antiproton, the particles have to be produced, collected into a beam and cooled so that it does not bounce around in the beam tube, and then steered (see Chap. 11).

In 1940, Enrico Fermi, perhaps on a whim, had proposed a 100 TeV machine that would be placed around the earth, without even anticipating advances such as the one in Tevatron, but he would be proud that the first machine to reach the TeV particle energy was built by a laboratory named after him. On September 30, 2011, the Tevatron collided its last particle and the eminent physicist Helen Edwards pressed the button to drain the particles from the machine was the last time.

HERA: Remarkable Marriage of Unequals

Perhaps the blessings from Hera, the Greek Goddess of marriage, was a part of this very successful and remarkable superconducting synchrotron Hadron Elektron Ring Alnalge (HERA) located in DESY, Hamburg, Germany. The same way, the first ever roofed temple to a Greek divinity was built for Hera, the first ever fully superconducting synchrotron was named HERA. The remarkable marriage is the

Fig. 8.11 HERA's intersecting electron/positron and proton rings. The detector in the straight sections is also shown

fact that this is the only collider which collides electrons and positrons with protons, so that one gets a combination of high energy as well as the relatively less messy collisions that come from colliding electrons or positrons. The proton beam nominal energy is about 920 GeV, close to a TeV, but the electron energy has to be kept low ~27 GeV to keep synchrotron radiation effects small. The proton ring (see Fig. 8.11) uses 422 superconducting dipoles which produce 4.7 T and 244 superconducting quadrupoles. A standard full 47 m long FODO cell consists of 4 dipoles, 4 quadrupoles, and 4 sextupole error correction magnets. The total circumference is about 6.4 km similar to the Tevatron. The injector ring PETRA with 2 km circumference has normal magnets. But the particles are privileged to be accelerated by a superconducting RF cavity which produced 5 MV/m. The electrons and protons are stored in separate bunches and then made to collide. In 1998, the machine switched to positron–proton collisions.

The hard work of technological development of not only superconducting accelerator magnets and RF cavities, but also of massive superconducting detector magnets has paid off. The days of superconducting synchrotrons with superconducting dipole, quadrupole, and field correction magnets and superconducting RF cavities have come and there is no turning back. The new higher temperature ceramic superconductors are already finding their place in leads, busbars, etc., and operation of the superconducting synchrotron with high stability at higher temperature may be in the future. Appropriately, the word "Super" has been co-opted and particle colliders are now "Super Colliders."

Chapter 9
Linear Accelerators, the Straight Story

When it concerns science, (a) No new serious idea, even when not accepted in the context when it was proposed, is ever wasted (b) the scientists do not give up easily.

This is the case with the linear accelerators. The straight accelerator was proposed and proven in the early 1930s after Rolf Wideroe encountered many hardships in implementing his idea of a linear (straight) accelerator (see Chap. 5) with a series of tubes, electrified alternately positive and negative to coincide with the arrival of the particles at the tubes. As particle energies increased, the machines got longer and expensive, less efficient, and the beam dynamics became more complicated. The cyclotron then showed that one could achieve high energies in a more compact machine and for a while, there was less interest in linear accelerators (LINACs).

Sloan Linac

Even so, in parallel, David Sloan and Lawrence agreed that LINACs were simpler and continued to reach for higher energies in them. Lawrence had spent two summers in 1929 and 1930 working on ionization phenomena. He and Sloan used a robust GE oscillator from those days, operating it aggressively at 11 kV where as it was rated for only a little over 2 kV and built a 90-keV accelerator with nine electrodes using Wideroe's concept. A 200-keV machine was then built and quoting this success, Lawrence, the king of circular accelerators, championed the cause of LINACs in his report to the National Academy of Sciences. In May 1931, Sloan crossed the one million volt milestone when he achieved 1.26 MeV mercury ions in a LINAC using 30 electrodes and 10 MHz RF source (Fig. 9.1). (It must be noted that the choice was limited to heavy ions which are slow for a given target energy, because high velocities meant high frequencies, high power, sources for which had not been developed). This was a phenomenal 42 kV per electrode and is impressive even by today's standards. The LINAC was instrumental in not only crossing the million volt barrier but also in providing experience with a great

R. Jayakumar, *Particle Accelerators, Colliders, and the Story of High Energy Physics*,   131
DOI 10.1007/978-3-642-22064-7_9, © Springer-Verlag Berlin Heidelberg 2012

Fig. 9.1 Sloan's 30-
electrode Linac, which
achieved 1.26 MeV mercury
ions, Sloan D.H. and
Lawrence E.O., (1931)*Phys.
Rev.* **38**, 2021.,Sloan D.H. and
Coate W.M., (1934), *Phys.
Rev.* **46**, 539

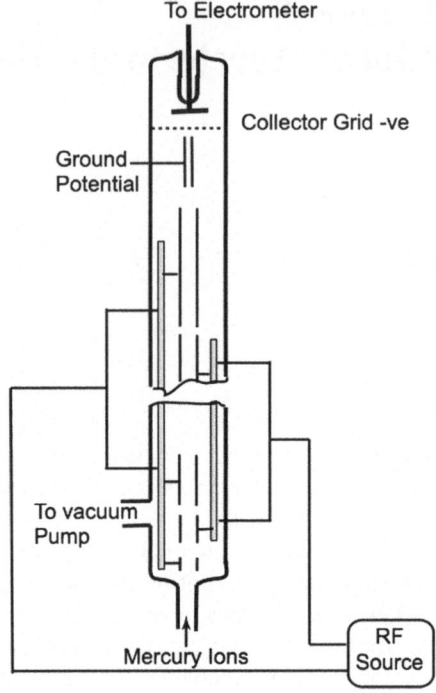

number of accelerator-related technologies like, vacuum, high voltage, detection
and measurement, etc. (Heilbron, J. L., and Robert W. Seidel *Lawrence and His
Laboratory: A History of the Lawrence Berkeley Laboratory, Volume I.* Berkeley:
University of California Press, c1989). With these devices, Sloan and Lawrence
addressed, for the first time, two important aspects of accelerator physics: the issue
of synchronization, that is, the arrival of particles in synchronism with the
oscillating voltage of the RF source and the issue of focusing.

In this arrangement, the particle is accelerated between the gaps of the electrode
tubes and less so inside the tubes because the field is shielded by the metal tube.
Alternate electrodes are at the same potential at any given instant and when the
particle arrives at a gap, the voltage should reverse on the tubes (half a cycle). At
non-relativistic speeds the time taken to travel between one midpoint of the gap to
the next by a particle with a velocity v, is L/v, L being the distance between these
two midpoints. The RF must go through half cycle in the same time. When the field
in the gap is in the wrong direction, the particle is nearly shielded from the field by
staying inside the "drift" tube. Since the RF frequency f is inverse of this period,

$$f = (1/2)(v/L) \quad \text{or} \quad L = (1/2)(v/f) \tag{9.1}$$

This can be written as

$$L = (1/2)(v/c)(c/f) = (1/2)\beta\lambda, \tag{9.2}$$

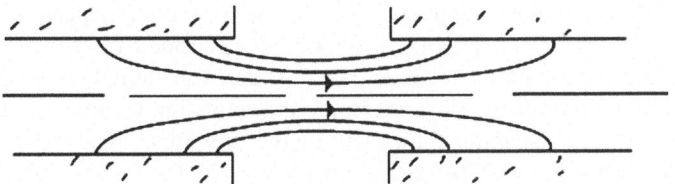

Fig. 9.2 Electric field (force) lines along which the potential gradients are established

where c is the speed of light, $\beta = v/c$ and λ is the wavelength of the RF wave. As one can see, the length of the tubes must increase in proportion with the speed of the particle.

Even if the maximum gap is limited to a liberal 0.25 m and a high frequency of 100 MHz is used (10 times used by Sloan), the maximum speed achievable is $v = 2Lf = 0.5 \times 10^8$ m/s $= 5 \times 10^7$ m/s, which is only about 16% of the speed of light. For an electron (mass $= 9.1 \times 10^{-31}$ kg) the corresponding kinetic energy $1/2 \, mv^2 = 0.5 \times 9.1 \times 10^{-31} \times 25 \times 10^{14} \sim 1.1 \times 10^{-15}$ J $\sim$7 keV, but a mercury ion, which is about 360,000 times heavier, could have an energy 2,500 MeV! This is the reason why ions were preferred. (But, of course, the final energy would be limited by the product of number of electrodes and RF voltage). The Sloan device also had an automatic focusing mechanism. When one calculates or measures the electric field lines (along with the accelerating voltages are established) between adjacent tube with the accelerating voltage difference, one sees that they are elliptical (see Fig. 9.2) convexing towards the axis. Since a particle can be thought of as being coaxed by the field lines to go along them – in analogy with iron filings along magnetic field lines – the particles would be pushed radially inward, causing radial focusing (There is a detail to it which makes the net effect to be focusing even though the lines open up as they flare out to the downstream electrode – see later).

Physicists were lured by the steady progress in physics and technology of circular accelerators and for a while LINACs did not make any significant progress. There was no perceived advantage in LINACs. The situation would change with the arrival of Luis Alvarez, an excellent physicist, but more, an inventor par excellence and a man who crisscrossed various fields ignoring boundaries between physics and other sciences.

Luis Alvarez: Renaissance Man

Commended by the American Journal of Physics as "...one of the most brilliant and productive experimental physicists of the twentieth century." Luis Alvarez was a man who planted his feet firmly at the University of California, Berkeley, but spread his intellectual wings and flew over a wide range of topics. With 40 patents, many of which are the foundations of the aircraft communication and radar technology such as Linear Dipole Array Antenna, Ground Control Approach for guiding airplanes for landing and take off, and short wavelength magnetron microwave sources, he dominated the scene of communication technology. He devised the microwave/radar apparatus for measuring the strength of atomic explosion from a plane. He co-discovered Tritium, invented the variable focus lens which every optometrist profits from, and invented a way to stabilize hand-held cameras. Very importantly too, he stamped his intellectual prowess in many fundamental physics topics ranging from Geiger Counter to observing inner shell electron capture thus demonstrating the accuracy of theory of beta-decay, to measuring magnetic moment of neutrons. He received practically every prestigious award for his various works and the Nobel Prize (1968) for the invention of the hydrogen Bubble Chamber (see later chapter). Luis Alvarez captured the imagination of common men by delving into socially intriguing topics. With an ingenious scheme using cosmic rays, he X-rayed Egyptian pyramids showing that many of them were empty and using photoanalysis, he showed that the Life Magazine's photographs of US President John Kennedy's assassination were inconsistent with the lone gunman theory of Lee Harvey Oswald being the lone assassin. Later in his life, in 1980 he and his son Walter proved that the extinction of dinosaurs could have come about because of an asteroid collision in the earth, which shook the scientific world and pointed out the future potential for a catastrophe. This brought him worldwide acclaim, well beyond his previous fame (Fig. 9.3).

This Leonardo da Vinci of our times also did a great deal to revive the interest in LINACs. Wolfgang Panofsky called him the "father of the modern practical proton linear accelerator". Alvarez believed that contrary to popular opinion, LINACs were more practical and less expensive devices than cyclotrons for high energies. He stated that for both, the dimension increased linearly with energy and he estimated that the cost of a circular machine would increase as the square to cube

Fig. 9.3 Luis Alvarez honored on the Republic of Guinea stamp of 2001

of the energy, whereas, LINAC cost would only increase linearly with energy. At this point, it is worthy to note that in the later years, the circular machines moved from being cyclotrons to synchrotrons, which are, in a way, essentially linear machines placed in a circular arrangement with additional bending magnets and do not have to meet the requirement of cyclotron resonance.

Interest in LINACs started increasing, when due to the efforts of people like Alvarez, high-power microwave sources even in GHz frequency range became available. After completing his work on the Manhattan Project at Los Alamos, New Mexico, and after World War II, Alvarez returned to Berkeley in 1945. Inspired by the renewed interest in LINACs, he assembled a team to build a 40-foot proton accelerator. The energy was not impressive even in those days, but would serve as a pilot project. Alvarez proposed using surplus military hardware so that costs would be kept low [Discovering Alvarez: selected works of Luis W. Alvarez, Ed. Peter Trower, The University of Chicago Press, Chicago, USA (1987)]. Alvarez and his team made several trade-off studies with respect to the total accelerator length for the target energy (or vice versa), RF voltage, and frequency. One other consideration was the limitation on the electric field (voltage per unit length) of about 2–3 MV/m, above which there was a strong possibility of emission of electrons and X-rays from the electrode surfaces, causing electrical breakdowns (The accelerator community has since learnt many lessons and in modern accelerators. As we saw in the previous chapter, it is possible to increase this limit by more than an order of magnitude).

In Wideroe's design one can think of the alternating voltage on an electrode as a wave of electric field passing over the electrodes. In vacuum such a wave would move at the speed of light, but the wave is loaded down in the structure and slowed down to match the particle's speed, so that the particle always sees an accelerating voltage (like riding a wave). But when the cavity is loaded, slow-down of the wave is accompanied by a large power loss. For electrons that have large speeds for a given energy, the losses are not heavy. But, in Alvarez's machine, even for the aggressive injection energy of 4 MeV from a Van de Graf generator, the speed of the proton beam is only about 9.2% of the speed of light and slowing down the wave to this low speed would entail substantial losses. LINACs are more effective (shorter) for higher frequencies, but at high frequencies the accelerating structure acts as an antenna and radiates considerable power. Therefore, Alvarez came up with a creatively new way of constructing the LINAC. This is shown in Fig. 9.4.

Instead of a traveling wave structure of Widereo, Alvarez arranged the tubes, termed "drift" tubes, such that they were inside a wider metal cylinder. In that case the electromagnetic wave of the RF field travels from one end to the other, gets reflected and forms a standing wave, which crests and ebbs at each point and the maxima (antinodes) and minima (nodes) locations are fixed. The physics of such a cavity results in an electric field which is mainly axial and the associated magnetic field is radial. In this mode, each end of the Linac tube has opposite polarity due to induction and the net charge is zero. The adjacent ends of adjacent tubes also have opposite polarity providing an accelerating gap. Each of these polarities alternate in

Fig. 9.4 Schematic of the
Alvarez Linear Accelerator
(Drift Tube Linac)

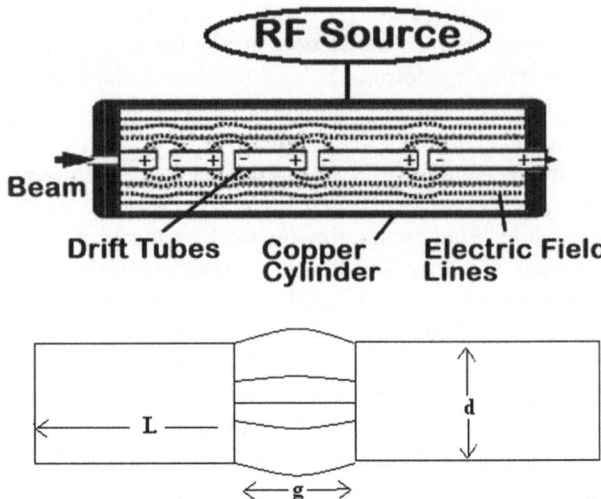

time corresponding to the frequency. In such a machine, Alvarez, Panofsky and team showed that the momentum of the particle increases linearly with the number of tubes. In the so-called π-mode of the device, there is a net current flowing through the electrodes which flows along the outer cylinder wall and through the RF connection to the tubes. In this mode, the outer wall becomes an effective shield and power is not radiated as in Wideroe Linac.

There were many complications in this new design. Clever rethinking was required to solve the associated physics problems and fructify this concept into an actual machine. Alvarez was unique in his knowledge and technical experience on RF physics and technology as well as in beam dynamics. He and his team addressed each of these issues and arrived at various trade offs. The first of these was the distance between the midpoints of the gaps. Since each tube itself has opposite charges at the two ends, each gap is accelerating and therefore the distance L between the midpoints has to be equal to $\beta\lambda$ rather than $(1/2)\beta\lambda$ as in Wideroe's machine.

Another complication in the machine is the requirement that the cavity has to be in tune (resonant) with the single frequency of the applied field at each tube (In Wideroe/Sloan design, the length of the tubes and gaps had only to increase with energy). Measuring and modeling showed that in order to get the resonance, the following condition was required to be satisfied (λ is the RF wavelength, about 1.5 m for this machine) (Discovering Alvarez: selected works of Luis W. Alvarez, See above)

$$g/L = -1.271 + 1.63(D/\lambda) + 1.096(L/\lambda) + 3.58(d/\lambda) \tag{9.3}$$

where d is the diameter of the tubes, g is the gap, D is the inner diameter of the large cylinder, and L is the length of the tube (Fig. 9.4). For a simple geometry which assures good axial electric fields, the gap g has to be small and fixed. But, as the particle acquires higher velocity (higher β), L has to proportionally increase to remain synchronous with the field, decreasing g/L. Therefore, d, the diameters of the successive tubes, have to decrease to compensate for the decrease in g/L and the increase in L/λ. So the tradeoffs on the parameters g and D had to be made so that at the peak energy, d does not become impractically small.

Yet another complication is that in such an arrangement, the electric field lines, though more or less axial, tend to concave away from the axis near the drift tube walls at the gap, unlike the focusing field lines of Sloan's machine and the focusing in the radial direction is not automatic. The phase or longitudinal focusing with slow particles being accelerated more and fast particles being accelerated less, can also be applied in this type of device, but the radial focusing conditions and the longitudinal focusing conditions turn out to be incompatible. While there is a small window in which both longitudinal and radial focusing can be achieved, such RF cavities are essentially radially defocusing (A fundamental theorem on this was published by McMillan). This is because, in crossing a gap the speed of the particle increases and therefore, the particle spends more time in the region with field lines flaring out radially, which is defocusing (left half of the gap in figure) than in the second half of the gap where fields return towards axis, that is, are focusing. The variation of the electric field during the transit of the particle in the gap also increases this effect. The only way to circumvent this problem was to introduce additional charges in the system. As he typically said, "it occurred to" Luis Alvarez that placing thin foils or metal grids in the gap, which would pass the particles and yet are a constant potential surface, would provide a radial focusing arrangement. This addition, though effective, required recalculation of all beam dynamics conditions and each of which was carried out diligently. Unrelated to this, the specific electric field values and the operating pressure conspired to create the problem of "multipactoring," an ionization phenomena, which occurs in specific field and vacuum conditions and can be easily detected but not cured easily. It was determined that applying a D.C. voltage on the drift tubes made the electric fields asymmetric and avoided the problem. But actual implementation of this idea turned into a nightmare of engineering and material problems, because insulators would break, vacuum leaks would occur, etc. But these were eventually solved.

The Alvarez proton accelerator was constructed in a 40 ft (13 m), 48.5 in. (about 1.25 m) diameter, 1.25 inch (about 3 cm) thick vacuum tank. A high conductivity copper liner, 38.5 in. (somewhat less than a meter) diameter, provided the outer wall of the cavity. There were 46 drift tubes, the first 11 with a constant diameter of 4.75 in. (about 12 cm) and the remaining 35 had decreasing diameters with last one having a diameter of 2.75 in. (about 7 cm). A grid holder held by a flange is attached to each end. Since the resonant tuning is critical and parasitic modes had to be avoided, end tuners (movable tubular inserts) were added. After the usual set of near-disasters and electrical, vacuum and mechanical failures in the Van de Graf injector and the accelerator, the machine was started in 1946 and within a few days, the full energy of 32 MeV was reached. Once the problems were understood, the reliability and predictability of the Alvarez machine was found to be higher than that of cyclotrons. Beam currents of 1 μA (impressive by any standards) could be achieved reliably. With this success, Lawrence proposed using the LINAC for breeding Pu and U isotopes.

The success of the Alvarez machine, established the science and technology of LINACs and the only thing that would revise its design for future accelerators would be the use of quadrupole magnets (see previous chapter) for focusing. Following this machine, John Williams and his team built a 60-MeV accelerator at the University of Minnesota which was used for proton scattering studies. The Alvarez accelerator was eventually moved to University of Southern California in Los Angeles, since Berkeley had moved on to the 184-in. cyclotron. In the late 1960s, Vladimir Teplyakov led a team at the Institute of High Energy Physics in Protvino, Russia, to build a 100-MeV Alvarez Linac. In 1972, a powerful LINAC with 800 MeV, high intensity proton beams was established as the Los Alamos Meson Physics Facility and continues to operate.

The Electric Analog of Alternate Gradient: RF Quadrupole Accelerator

In 1969, Teplyakov came up with a Radio Frequency Quadrupole concept for a LINAC [I.M. Kapinchskii and V.A Teplyakov, *"Linear Ion Accelerator with Spatially Homogeneous Strong Focusing"*, Pribory I Technika Eksperimenta (March 1970)], and in many ways, this design is even more insightful than the Alvarez design. It has been called the "missing link" in LINAC technology and shows the depth of knowledge and innovativeness of the inventor. This uses scalloped outer walls and creates alternately focusing and defocusing electric fields, just like the alternating (magnetic field) gradient quadrupoles used in synchrotrons (see Chap. 7). While the Alvarez Drift Tube Linac (DTL) could be used only with high injection energies, that is, $\beta > 0.04$, the RFQ is not limited by this requirement, because of the use of quadrupole electric fields that are always focusing (Fig. 9.5). In this geometry, four electrodes (vanes) are placed symmetrically and coaxial to the beam and are powered by an RF source, such that adjacent electrodes have opposite polarity (quadrupole arrangement). If everything is symmetric, no axial potential exists and a pure focusing field exists. But, if a pair of opposing rods are moved away or in, there is a non-zero accelerating electric field on axis. Instead of just displacing rods, if the rod surfaces are undulated along the length, then an axial accelerating electric field is created (see figure below – an alternative arrangement is four rods with diameters increasing and decreasing). The adjacent vanes are at opposite polarity (Fig. 9.6 – a simplified geometry) this gives a gradient of potential shown in the figure on the right.

A proton would be accelerated in the region where the potential is decreasing (left of the figure) (However, this accelerating field due to scalloping comes at a cost of reduced focusing). Like the Wideroe machine, by synchronizing the period of the RF field with the arrival of a particle at the accelerating location, the particle

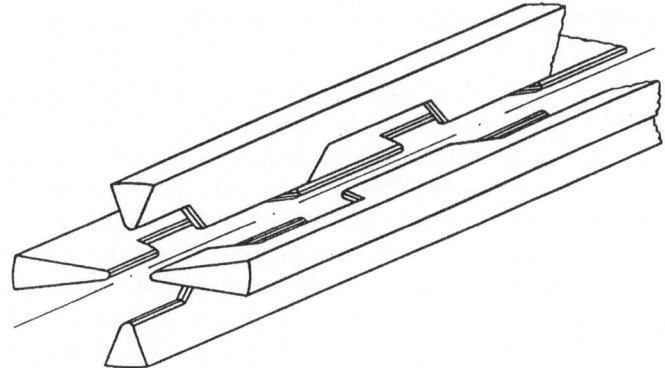

Fig. 9.5 Basic structure of RFQ electrodes

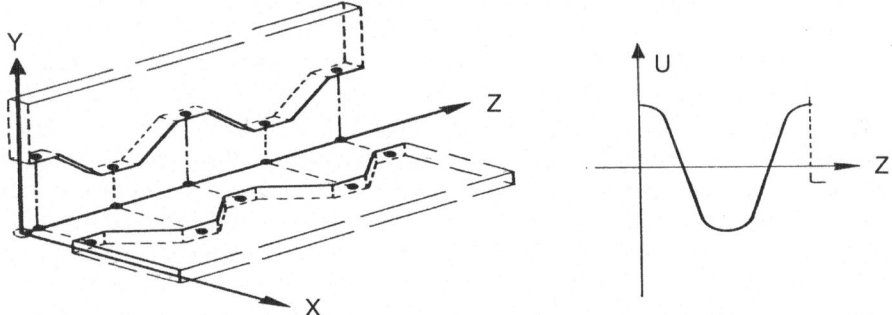

Fig. 9.6 Two electrodes transverse to each other (*left*) and the potential between them as a function of length (Courtesy: CERN, Eurpean Organization for Nuclear Research)

is accelerated and longitudinally focused in every other gap, and is radially focused in every other gap. This is reminiscent of the AG focusing in magnet-guided accelerators. The long length of focusing makes it superior to other LINAC designs.

Teplyakov received the highest award of Lenin Prize in Soviet Union and also was recognized by the American Institute of Physics, for this invention. This highly successful concept is still in use today as an injector for high-energy accelerators, often immediately following the ion source, because of its applicability even at low particle velocities. As stated above, high-energy Linacs, like the synchrotrons, use quadrupole magnets for strong focusing. Since the force due to a magnetic field is proportional to the particle velocity, such a scheme is not useful at low energies. Therefore, in the accelerator chain with stepped-up energies, the first LINAC is, more often than not, an RFQ Linac. Its compactness and ability to obtain high currents, recommends itself to many applications including focusing or cooling of beams, but the application has been limited to low energies or heavy ions. This is because while the particle momentum and therefore the current increases linearly with the particle speed, the length of the RFQ increases rapidly as cube of the velocity.

Stanford Linear Accelerator Center: 2-Mile Marathon

There is a symbiotic relationship between particle beams and radiofrequency waves of all wavelengths. Electron beams when modulated in velocity, would couple to radiofrequency waveguides, which would resonate, draw energy from the electron beam, and produce RF power. This class of RF and microwave devices – magnetrons and klystrons are bread and butter of the RF technology. In a different application, when this RF power is fed into appropriate structures of the LINACs described above, these produce the accelerating voltage to accelerate particles. For acceleration to a target energy of T, the RF power is the main constraint and is proportional to $T^2\lambda^{1/2}L$, λ being the RF wavelength and L being the length of accelerator. Therefore, one has to choose the best available highest frequency (lowest feasible wavelength) sources that can give highest voltages. The length, of course, is reduced by increasing the accelerating voltage. In practice, availability of sources, close mechanical tolerances, and electrical breakdowns moderate these decisions.

In 1935, two brothers Russell Varian, a former physics graduate student in Stanford University and Sigurd Varian, a pilot but with no particular technology expertise, set up a small company in Halcyon, CA. During his war service, Sigurd had realized that one needed high-frequency radio sources that could be focused like searchlights to search for air-raid planes and had persuaded his brother to start the effort for developing these. They sought the help of Bill Hansen, Russell's advisor and the facilities available at the Stanford University physics department. The University made a sweet deal in which, for a payment of mere $100 for materials and some space for the Varian brothers to work, the University would get half the royalties from any invention. Within just few months in which they exhausted many variations, their 23rd design incorporating some of the concepts developed by Bill Hansen, gave rise to a successful design of a klystron microwave source. Under the headline "Hitler Warns- Leave us alone," Palo Alto Times reported, "New Stanford invention heralds revolutionary changes." By 1940, the brothers had several commercial designs and these became important tools during the war.

Encouraged by these developments and inspired by the success of the Alvarez linear accelerator, Bill Hansen, who was a big believer in LINACs, built a 6-MeV electron accelerator using a 1-MW magnetron RF source. Following his terse four-word report – "We have accelerated electrons!", he submitted a bold proposal in March 1948 to the Office of Naval Research (ONR) for a 1-GeV electron LINAC, 160 ft (about 49 m) long, with a 30-MW klystron every 10 ft (No matter that this RF power was 2,000 times the power they had achieved so far). He even claimed that 1 GeV was a conservative estimate and with the expected improvements in klystron efficiency they could double the energy. An interesting point is made in this proposal: it states that the accuracy of the 160 ft, 0.75 in. diameter beam tube would not be a challenge, because the electrons would only think it is 6.75 in., because of relativistic Lorentz contraction (But in the direction transverse to the velocity the dimensions would look the same, allowing normal tolerances). In spite of the audacity of such a proposal and uncertainties in the program, the ONR saw this as important for military uses and accepted the proposal. The program started in 1948. The klystron development went well and Hansen and 29 grad students worked on the accelerator. Unfortunately, in 1949, even as the 6 MeV, Mark II electron accelerator (Fig. 9.7) was being built and preparations were being made for the 1-GeV accelerator, Bill Hansen developed a chronic lung illness and died. The death of this brilliant man, at a relatively young age, shocked the Department and the grieving team also worried how they would do without him. Edward Ginzton and Wulfgang "Pief" Panofsky ably stepped in and guided the development at this juncture. The team built a succession of prototype accelerators, Mark II-14 ft long (giving 33 MeV) and then in January 1952, Mark III-80 ft long (giving 200 MeV). Though the optimistic energy goal of Hansen was not met, in 1957, a 220-ft version of Mark III achieved 900 MeV regularly and 1 GeV when the klystrons occasionally worked at peak efficiency. By now, the accelerator facility had been named after Hansen. Between 1954 and 1957, Robert Hofstadter carried out experiments on scattering of electrons by nucleus and surmised the structure of protons and neutrons, using the Mark III accelerator. This brought him a Nobel Prize in 1961.

The total cost of Mark III, despite having delays and ups and downs, was under a million dollar and there was much enthusiasm for a 10-GeV machine at $10 million – a bargain (and a highly underestimated number). (Informal history of SLAC, Part 1, by Edward Ginzton, SLAC beam line, April 1983). A group of LINAC enthusiasts

Fig. 9.7 William Hansen and graduate students holding a 3 ft section of the Mark I linear acclerator Alvarez and his associates were seen sitting on the vacuum tank of their (large) machine. When he learned of this photograph, Hansen demonstrated that he could carry his machine, the Mark I linear accelerator, on his shoulder!- [Credit: Wolfgang K.H. Panofsky in the SLAC Beam Line, spring 1997, p. 41. Symmetry magazine, Vol. 2, Issue 6 (2005)]

met at the home of Panofsky in April 1956, and a year later, proposed to the Atomic Energy Commission that a LINAC project with 15 GeV initial target, expandable to 50 GeV, should be established with Edward Ginzton as the Director. The machine's maximum length would be 2 miles, longest stretch available within the University grounds. It would use 480 klystrons each of 6 MW. Early on, it was recognized that such a large machine would be a national facility like the Brookhaven machines and not like the Mark II and III, intended for the faculty and students. However, this machine was a maverick running against the popular current of circular machines. One drawback of LINACs is that while the beam is intense, it has a low duty cycle and the target collisions are very brief. This means that data has to be acquired and coincident events have to be resolved in time more accurately. The second drawback is that a forward shower of intense secondaries is created in such collisions, and particles of interest have to be identified in this haystack of debris. Therefore, any such facility would have to be staffed with top-notch experimentalists to build and operate these sophisticated detector instruments.

As is common even nowadays, there were fears about the faculty and the department being devoured by this "Monster" project and academic freedom would be jeopardized if this facility, now named Stanford Linear Accelerator (SLAC), was inside the Department like the Hansen Laboratories. Therefore a special arrangement was made. Panofsky (Panofsky, W. K. H. An Informal History of SLAC Part Two: The Evolution of SLAC and Its Program. SLAC Beam Line Special Issue Number 3, May 1983) states, "... administratively SLAC would be entirely separate and would thus not drown the existing administrative machinery of the University. SLAC would operate under a general policy set by the University, but its actual operation would be almost autonomous...". In 1959, President Eisenhower expressed favorable opinions and said that the project would be supervised by AEC. While the technical design of the machine progressed well and congress received a four-year study in fall 1961, the project was held hostage to political horse trading between the congress and the President. During a congressional hearing, when a Senator asked *"Dr. Ginzton, can you tell me exactly why you want to build this machine?"* Ginzton gave a snappy but quirky reply *"Senator, if I knew the answer to that question, we would not be building this machine"* (If all this seems familiar, then indeed it is, because this happened with the birth of Fermi National Laboratory. It appears that politicians who vote on science projects, never read the history of science and never tire of this question. So, it falls upon the protagonists to repeat their answers and come up with catchy phrases).

The SLAC cavity was different from the Alvarez Linac. The disadvantage of previous Linacs was that in the time the particle spent inside the tube, it was not being accelerated and as the energy increased, this wasted length increased. Therefore, a design, in which nearly the whole length of the machine is used for acceleration, was needed to reach high energies. One way would be "load" the structure (cavity) so as to reduce and match the phase velocity of the wave to the particle velocity with the so-called "disc loaded waveguides" (Fig. 9.8) (see Accelerator Physics by Shyh-Yuan Lee, World Scientific Publishing, Singapore (2004)). It must be noted that such machines may be used in the standing wave

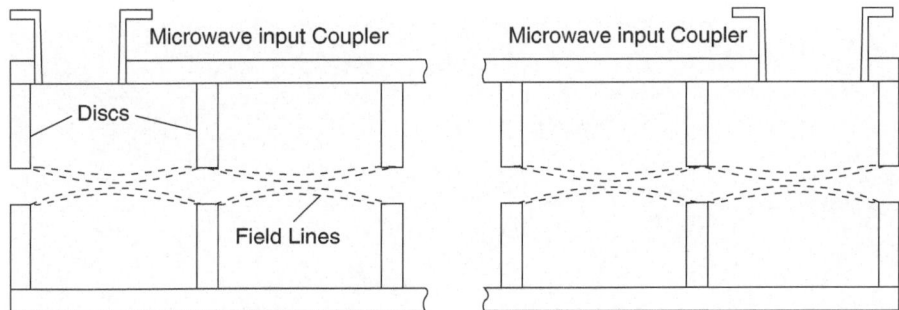

Fig. 9.8 Schematic of Disc-loaded Waveguides used in SLAC; *dotted lines* show the accelerating electric field lines

mode as Alvarez Linac or traveling wave mode (In the standing wave mode in which a wave travels in one direction and reflects at the end and travels back, the wave that is going in the direction of the particle beam is the one that provides most of the acceleration of the beam). The SLAC machine uses mostly a traveling wave mode, which is considered to be more appropriate for pulsed operation. The SLAC design also ensures that the particles see a constant accelerating gradient.

So, in the SLAC machine, metal discs with holes instead of thin wall tubes were mounted normal to the outer structure with their central hole allowing passage of the beam. The high-frequency RF is coupled periodically at several points in the section and the discs couple to the wave through the central hole (like mechanical structure resonating to a sound wave). By adjusting the dimensions of the disc, one can tune the structure to resonate with the applied RF. As the wave travels down the structures, the dimensions have to be adjusted to account for the power loss in the wave and keep a constant accelerating gradient along the length. However, radially the gradient is not constant and is non-uniform, but has a focusing characteristic. While this structure fabrication is much simpler than the tubes, the physics of these waveguides is very complicated and require sophisticated analyses. For electron accelerators, it becomes somewhat easier because the electron speed reaches near-electromagnetic wave (light speed) velocities at 0.5 MeV.

One of the strong points of LINAC is that unlike circular accelerators, where the machine is considered to be working only if the particle does many orbits with high intensity in a fully commissioned accelerator ring, a successful section of LINAC is a necessary and sufficient proof of the viability of the design, construction, and operation. A LINAC is just a string of these sections. This means that a section itself becomes a functioning prototype providing proof of principle and this prototype would become part of the larger machine. Therefore, in terms of risk, the escalation of the particle energy to 30 times the prevailing energy did not pose a serious risk. Therefore, for the SLAC, only a 800 ft section was made first and the 2-mile machine was gradually finished (Fig. 9.9). New fabrication methods were devised and a lot of material processing work like machining, annealing, polishing, brazing, and finishing of the disks and spacers were done by housewives looking for part time work. Their meticulous work showed its quality when none of the brazings

Fig. 9.9 2 mile long linear accelerator (the straight line heading to top right), at SLAC. Photographed in 2006. The figure also shows a circular machine PEP II in the foreground. The linear acceleartor injected electrons and positrons into two rings in PEP II, which were then collided. (Courtesy: SLAC National Accelerator Laboratory)

leaked under vacuum even after two decades of operation. After several collaborative attempts with industry for meeting the demanding specifications of klystrons, SLAC finally had to develop the klystrons themselves and did so successfully and achieved 20,000 h of operation without failure.

One of the unusual details that had to be taken care of for this 2-mile long accelerator was that it was being located in earthquake-prone region and the required precision for mechanical stability was great. The civil engineering work done in this regard was exemplary and led the way for building earthquake-proof accelerators and colliders of today. The civil work was carried out in the open Stanford grounds and an amusing anecdote is related by Panofsky.

Feeding power from the klystrons to the accelerator required very complex waveguide plumbing. We decided to mock up a prototype consisting of a single klystron feeding an accelerator section through the actual waveguide system. To provide for adequate shielding from the Linac, the klystrons must be 25 ft above the machine in the actual installation. Therefore, this mockup had to be constructed as a tower which contained the klystron and its supplies while the accelerator section was placed at ground level. The easiest way to install the waveguide feeds from the upper story of the tower down to the accelerator was by helicopter (a method later used in the actual accelerator construction). As it happened, this mockup tower was next to the Stanford football stadium, and it also happened that the lowering of the waveguide by helicopter was made on the Friday before a critical game. The Stanford football coach was practicing some very secret formations in the stadium at the time and thought the helicopter was part of a spy operation by Saturday's opponent! He canceled the practice, and when informed of the actual situation sent a strong letter of protest to SLAC.

The 2-mile long machine, occupying 480 acres and costing $105 million was commissioned with relatively small difficulty, after requiring further alignment of section to prevent beam losses due to scattering from the RF wave reflected from the exit end. In early 1967, the machine exceeded the target energy and obtained 20 GeV electrons, with a current of up to 300 μA for 1.7 μs. The advantage of high-energy electron accelerators is realized in obtaining a narrow cone of secondary particles in the forward direction from a target collision and also a near-monochromatic beam of X-rays which are extremely useful for science as well as technology. The SLAC also became the first machine to produce abundant neutral K-mesons without producing other obscuring particles like neutrons. Study of these Kaons was instrumental in understanding symmetries in weak interactions.

LINACs are important as injectors for circular accelerators and for accelerating electrons, since in circular machines, the electrons would lose their energy due to synchrotron radiation during their bending. These even remain as contenders for a broad range of accelerators because, depending upon the project needs and technological innovations in microwave technology and RF cavities, they can become competitive to circular machines. A 27–40-km long 500 Gev International Linear Collider has been proposed and being designed by an international team, coordinated at the SLAC. These beams are expected to provide intense collisions with higher measurement accuracy than more energetic circular accelerators. Like in social morality, the straight and narrow path is advocated and is always a good option.

Chapter 10
The Lotus Posture, Symmetry, Gauge Theories, and the Standard Model

Since the beginning of physics, symmetry considerations have provided us with an extremely powerful and useful tool in our effort to understand nature. Gradually, they have become the backbone of our theoretical formulation of physical laws.

T.D. Lee

In the Lotus posture of Yoga, as in praying posture in other religions, where oneness with the Universe is the goal, one sits in a symmetric posture and fixes the mind on nothingness or oneness. Symmetries are deemed to be "principles of simplicity." This is symbolic of the deep human understanding and intuition that a disturbance of symmetry creates action and makes us take note of the diversity of the world around us. Quantum mechanics and eastern thought even indicate that reality is created when we break symmetry and apply our senses and sensors. But beyond this heuristic appreciation of symmetry, we are aware that breaking symmetry is also what we see as part of the beauty in our world. Although leaves on a tree are identical, they are so only to a point. They are slightly asymmetric and slightly different from each other creating the rich mosaic of diverse tree leaves and branches which make up the beautiful tree. So it is with particle physics, it is well understood that if everything was symmetric in the physical description of the Universe, the Universe would be a quiet place in silent repose, like the Hindu God, Narayana in His Ocean of milk. While some of this symmetry is evident in very understandable ways, most of the symmetry that underlies the rich diversity of particles, forces and fields can only be seen in mathematical terms, and are revealed through physical laws. Yet, breaking or violation of these symmetries is also the reason for the myriad phenomena and forms of matter. Since the 1960s, there have been extraordinary developments in theoretical physics, some beyond the grasp of even reasonably trained mathematicians and physicists. But, a minimum conceptual understanding of the symmetries and associated forces controlling particle behavior can be garnered to understand these extraordinary developments. Such an understanding is valuable in appreciating the need for the modern day colliders and detectors.

R. Jayakumar, *Particle Accelerators, Colliders, and the Story of High Energy Physics*, DOI 10.1007/978-3-642-22064-7_10, © Springer-Verlag Berlin Heidelberg 2012

In this glimpse of the underlying physics, one might at times feel like Alice in Wonderland on the other side of the mirror, developing an understanding through what seems very strange but vaguely familiar and at other times seeing stark realities. The present-day description of High Energy Physics includes the following: (a) Different types of Symmetries in nature, some well known to us and some beyond the reach of our imagination and obtained through observing mathematical symbolism; transformations that demonstrate these symmetries; breaking of these symmetries which "precipitates" and differentiates different forces and interaction of particles with these forces; (b) Conservation Laws that flow from these symmetries, some which may seem trivial at first and taken for granted by us and others which are quite surprising, violation of these conservation laws that manifest the breaking of symmetry; (c) The mathematical description of the symmetries in "Gauge" Theory, specific to the description of a target physics area (for example, electromagnetism).

Symmetries and Transformations

A German mathematician Emmy Noether proved a theorem in which she demonstrated the fundamental justification for conservation laws, which shows that conservation laws follow from the symmetry properties of nature. This arises out of deep mathematical connections between different laws that govern mechanics, forces, and fields. A very rough idea can be given by the example of a round balloon which is blown up. The balloon remains symmetric and round once it has expanded. However, if the air is escaping and the quantity of air is not conserved, it becomes tear-shaped and is no more symmetric. To give a glimpse of basic physics – take the case of an object which is moved from one place to another and then moved again from there to another place. The fact that we might make these moves not now but a day later (with nothing else changing) (translation in time) without affecting the result (symmetry in time), results in the Law of Conservation of Energy. Similarly, imagine a particle moving from a start to an intermediate and then a final location. Now if all these locations are moved by the same distance in the same direction (all shifted together) and no additional forces are acting at this new location, then the particle travels the same distance and takes the same time. This symmetry leads to the law of conservation of momentum. Each symmetry property is complementary to a specific property of the particle. There are more symmetries in nature, many of them highly conceptual, discernible only through mathematical relationships and described only through very approximate analogies. In applying these symmetries to particles, the particles have associated parameters (like parameter C for charge conjugation, P for Parity) that one can follow and trace through a process or a phenomenon to see if this parameter is conserved.

It is more understandable that conservation laws are describing invariant quantities, so that when we say energy is conserved, we mean that when a certain phenomenon occurs or when a certain transformation is carried out, the energy remains invariant (in this context, a particle process such as a decay or a reaction may be considered to be transformation). We know that the number of pebbles in a collection is invariant, because in whatever order we count, we get the same count. Distance between two ends of the ruler is the same irrespective of where it is. However, equations describing relationship between properties need also to be invariant under a certain scheme depicting a process or phenomenon or transformation. Newton's laws of mechanics are invariant under rotation and translation. In Einstein Theory of Relativity, the quantity $E^2 - p^2c^2$, where E is the total relativistic energy, p is the relativistic momentum and c is the speed of light, is invariant, that is, independent of the coordinate system (a fact used in Chap. 11 to derive the energy available in beam-fixed target collisions). In quantum mechanics, angular momentum is invariant. Facts like the distance is invariant when a car travels with the same speed in the same duration independent of what the exact time is (other things like air resistance being the same), are generally applicable. The fact that all directions are equal (homogeneity of space) and the behavior is the same independent of the origin of reference is another symmetry we vaguely understand. The geometrical invariance that is difficult to grasp is that, phenomena within a frame remain the same independent of the velocity with which the frame is moving. Although we experience this in a moving train, the concept remains nebulous in our mind, and the relativistic consequences of these boggle our mind (Galileo had a tough time arguing that one could not determine the motion of the earth by observing motion of objects on the earth). These are the invariants, associated with symmetries that enable us to make sense of the world around us.

The general principle is that a property of a system is invariant if it remains the same after a transformation, and this invariance reflects a specific symmetry. A shape of a smooth sphere remains invariant under a rotation, which points to rotational symmetry. The relatively better understood geometrical symmetries and invariants are, in physics, supplemented by "dynamical" invariants. The dynamical invariants are only applicable to specific interaction or phenomenon. One can have other specific symmetries, such as charge conjugation: If a particle's charge is reversed and all relevant properties are also reversed, then the particle has C-symmetry. For example, if an electron's charge is reversed and correspondingly the electric and magnetic fields are also reversed, then the new particle (positron) would behave (actually behaves) exactly like the electron would in its environment. (Charge conjugation also involves reversing internal quantum numbers). This arises from laws of electromagnetism. If this were not so, it would have been a violation of the C symmetry. A neutral pion which decays into photons through electroweak process has a C parameter of +1. C is conserved only with a decay by the emission of two photons since the photon has the C parameter of -1 (the net parameter is the product of each). In particle physics a concept that had caused a lot of consternation but brought a new understanding, was the concept of left–right (or top–down) symmetry (mirror symmetry), which is called parity (P).

Often, in physics, especially in quantum mechanics, the properties of an object or a particle are a "state" of a particle described by a set of vectors, for example, the x, y and z positions or x, y and z components of momentum. A change of that state such as a rotation or reflection would be represented by a so-called operator, which often would be a matrix. Now if a particle goes from state a to state b first through an operation P acting on it and then goes to c through the action of the operation Q, then this would be written as $c = Q(Pa)$. Now if the same process is repeated

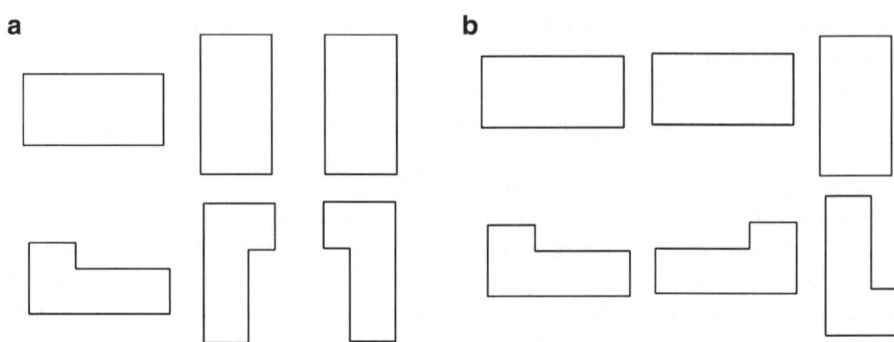

Fig. 10.1 Symmetry and commutation of operation. (**a**) The object is first rotated by 90° and then reflected horizontally (*left to right*), (**b**) the object is reflected horizontally and then rotated. For the symmetric rectangle (*top*), the operation is commutative – both the orders of operation give the same end result; but not for the polygon (*bottom*)

differently by applying Q operator on a and P later, then the state would be described by $P(Qa)$. Now if this final state is also c, i.e. $c = P(Qa)$, then we can write $PQ = QP$, and the operators are said to commute. This commutability indicates an invariant nature of operators. For example, if one of the operators P or Q is rotation, then this indicates conservation of angular momentum – if a system is the same irrespective of the stage in which rotation occurs, the angular momentum is conserved. This conservation is then described by the symmetry with respect to angle, namely isotropy. Therefore, symmetries, conservation laws and invariants are understood and analyzed through transformational manipulations. An example of symmetry exhibited by commutation of operation is shown in Fig. 10.1. The rectangular has an up–down, left–right symmetry and the operation of 90° rotation and a reflection about a vertical line can be done in any order. But the result of the operations for the polygon will depend on the order in which this is done, exhibiting a lack of symmetry in these directions.

The case of symmetry in charge conjugation and parity opened up new understandings in physics. A range of similar symmetries and breaking of these symmetries manifests or characterizes different properties of particles such as the charge, mass, and the forces they experience. In this approach, physicists postulate certain symmetries; then by examining the equations and transformations in that light, they arrive at plausible theories.

Violation of C, P, T Symmetries and Discovery of the Force of Weak Interaction

When equations of electromagnetism and gravitation are examined under the above process of transformations, and associated phenomena studied, one finds that these fields and forces conserve associated quantities. Certain properties may be expected to be invariant – quantities associated with charge conjugation (*C*), Parity *P*, (spatial reflective symmetry) and Time-reversal (*T*), where changing the direction of time does not change the particle or field behavior. Charge conjugation simply refers to change of a particle into its antiparticle and its symmetry is the equivalency of particle and antiparticle properties. A C operation on neutrino shows that lack of C symmetry for the neutrino. A neutrino is left handed and a Charge conjugation will result in a left handed antineutrino. But all antineutrinos are right handed. An example of these changes is shown in Fig. 10.2.

In each case of reversal, if the original and the transformed (reversed) outcome are both observed, then the quantity is deemed to be conserved. For example, *CP* is conserved in electromagnetic interactions because electrons and positrons have corresponding properties in an electric field (when the particles are conjugated and the coordinates are also reversed, they behave the same). Till recently, strong forces were also believed to conserve these quantities (and have additional symmetries specific to strong force interactions). But, as noted in an earlier chapter, during the tumultuous period of late 1950s and early 1960s, there was a proliferation of mesons which did not conform to any organized arrangement, and there was also not a good understanding of how and why nuclei and some particles decayed. The discovery of violation of the parity conservation was the clue to tame these mesons and explain the weak interaction force that was identified to be the cause of radioactive decays (The term violation does not mean that somehow nature's laws are violated, it simply means violation of earlier notions on conserved quantities). The story of the discovery is the stuff of history, and has been told many times in personal as well as scientific terms [See, for example, Women at the edge of discovery: 40 true science adventures by Kendall F. Haven, Libraries Unlimited, Westport, CT (2002) and The God Particle by Leon Lederman and Dick Teresi, Dell Publishing, NY (1993)]. The description below presents only the brief history.

Neutron Decay	n ->	$\bar{v}$	p	e⁻	
		<--		-->	Momentum Direction
	<--	<--	-->	<--	Spin Direction
Time(T) Reversal					
	n ->	$\bar{v}$	p	e⁻	
		-->		<--	Momentum Direction
	-->	-->	<--	-->	Spin Direction
Parity(P) & Time(T) Reversal					
	n ->	$\bar{v}$	p	e⁻	
		<--		-->	Momentum Direction
	-->	-->	<--	-->	Spin Direction
Parity(P) & Time(T) and Charge Conj Reversal					
	$\bar{n}$ ->	v	$\bar{p}$	$\bar{e}$+	
		<--		-->	Momentum Direction
	-->	-->	<--	-->	Spin Direction

Fig. 10.2 A flow of changes by *CPT* reversals from observed neutron decay into an antineutrino, a proton and an electron, to the surmised decay of antineutron into a neutrino, an antiproton and an antielectron (positron)

Parity (*P*) Violation

Most transformations that we can readily understand such as rotation and translation can be made incrementally. But there is one operation that is done in one swoop
– reflection. This transformation property is notable because particles have a
property called chirality (left handedness or right handedness; for example, we
can only digest Dextrose, the right-handed sugar molecule). When a transformation
preserves chirality, it is called parity. However, since parity is being now used to
represent chirality itself we will also follow this terminology. Parity reversal
reverses the sign of the position vectors but axial (a line about which reflection is
made) vectors remain the same. So, in a parity reversal where all the three position
vectors are reversed (x to $-x$, y to $-y$, z to $-z$), linear momentum would change
direction, angular momentum would not be reversed. Note that this is not mirror
symmetry as we experience it with a single mirror, because we normally experience
only left–right reflection.

   In the 1950s, two particles were discovered, the positively charged tau particle
(not to be confused with the present-day designation of a lepton particle) by C.F.
Powell. The tau decayed into three pions (two π^+ and one π^-). Then a "theta"
particle, which decayed into two pions (π^+, π^0), was also discovered. But both these
particles had identical masses, same spin and same scattering properties. Physicists
first thought that these were the same particles, just having two decay modes (the
decay of the tau and theta, now called Kaons, is through the force of weak
interaction). But, in 1953, the Australian born Oxford physicist R. H. Dalitz argued
that since the pion has parity of -1, two pions would combine to produce a net
parity of $(-1)(-1) = +1$, and three pions would combine to have total parity of
$(-1)(-1)(-1) = -1$. Now, if parity is conserved, the theta should have parity of
$+1$, and the tau of -1. Hence, they could not be the same particle (Trigg, G. L.,
Disproof of a Conservation Law, in Landmark Experiments in Twentieth Century
Physics, 1975). Many physicists were skeptical that two different particles could
have such identical properties and differ only in parity and the decay mode. There
was a raging debate. Two theorists, T.D. Lee and C.N. Yang dedicated themselves
to the investigation of this issue while visiting the Brookhaven National Lab. One
proposal they came up with was that, these "Strange" particles (the then prevailing
definition – heavy particles that were created frequently in interactions and yet,
"strangely," had relatively long life times given their large mass) came as a pair – a
"parity doublet." After hearing the talk, Martin Block boldly suggested to the
famous physicist Richard Feynman that perhaps these two were identical particles,
and one of the decays did not conserve parity. Though this was close to heresy,
Feynman thought that this was an exciting idea, but bet against it. On the other
hand, Eugene Wigner, who actually formulated the law of conservation of parity,
also agreed with the possibility that parity might be violated.

   In a landmark discussion between Yang and Lee, they decided to look into the
possibility by looking into all the experimental results. When they did an exhaustive
and careful analysis befitting the objective approach of true physicists, to their

amazement, they found that there had been no results that unequivocally pointed to conservation of parity in weak interactions, governed by the "weak" force. Then they went on to publish a paper on the topic in June 1956 and proposed experiments to confirm whether parity is conserved in weak interactions. One of these experiments is as follows: radioactive cobalt (Co^{60}) would be placed in a magnetic field so that there was a preferred direction for the nuclei, which have magnetic moments and so align themselves with the direction of the magnetic field. As these nuclei spun, beta rays (electrons) would be emitted from the "poles" (along the axis of the spin) of these nuclei (actually, this is a simplification – in reality, they have an angular distribution). Now, if parity was conserved then the two poles would emit same amount of rays (symmetric) in opposite directions (Fig. 10.3). However, if the two poles emitted different amounts, then one would be able to distinguish the reflection from the original (in essence, be able to tell from the direction of the magnetic field, which pole would emit more particles), and parity would be violated.

Lee and Yang proposed a similar experiment involving decay of muons into electrons and neutrinos (by hitting them on a target). If the parity is violated, again there would be asymmetric emission. Hearing about this proposal and excited by the physics intrigue, C.S. Wu cancelled her trip to China and started performing the first experiment in Columbia University, even before Lee and Yang's paper was published. Though simple in concept, the experiment was tremendously difficult. Wu and her team went through many difficulties, typical of these kinds of experiments – obtaining stability, suppressing thermal effects by cooling the experiments down, etc. Finally, on January 9, 1957, Madam Wu's team concluded that there was clear evidence that parity was indeed not conserved. The beta rays were being emitted preferentially in the direction opposite to the spin vector of the nucleus. This was earth shattering news. Such a result required multiple confirmations.

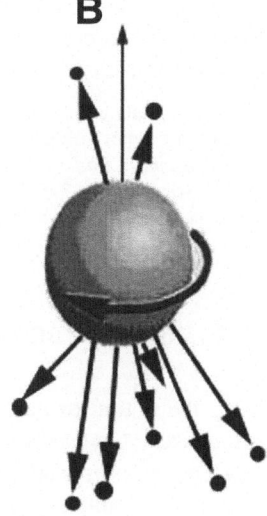

Fig. 10.3 When placed in a magnetic field, the radioactive decay of Co-60 results in a preferred direction of beta ray (electron) emission

Madam Wu immediately communicated this to fellow Columbia experimentalist Leon Lederman (a true handyman and later Nobel Prize winner and Director of the Fermilab), who had been dodging Lee and Yang about doing the second experiment. Lederman now jumped on this experiment, using the muon beam being created in a source built by Richard Garwin. Here is what happened.

Intrigued by the experiments of Madame Chien-Shiung Wu, Lederman called his friend, Richard Garwin, to propose an experiment that would detect parity violation in the decay of the pi meson particle. That evening in January 1957, Lederman and Garwin raced to Columbia's Nevis laboratory and immediately began rearranging a graduate student's experiment into one they could use. "It was 6 p.m. on a Friday, and without explanation, we took the student's experiment apart," Lederman later recalled in an interview. "He started crying, as he should have."

The men knew they were onto something big. "We had an idea and we wanted to make it work as quickly as we could – we didn't look at niceties," Lederman said. And, indeed, niceties were overlooked. A coffee can supported a wooden cutting board, on which rested a Lucite cylinder cut from an orange juice bottle. A can of Coca-Cola propped up a device for counting electron emissions, and Scotch tape held it all together.

"Without the Swiss Army Knife, we would've been hopeless," Lederman said. "That was our primary tool."

Their first attempt, at 2 a.m., showed parity violation the instant before the Lucite cylinder – wrapped with wires to generate the magnetic field – melted.

"We had the effect, but it went away when the instrument broke," Lederman said. "We spent hours and hours fixing and rearranging the experiment. In due course, we got the thing going, we got the effect back, and it was an enormous effect. By six o'clock in the morning, we were able to call people and tell them that the laws of parity violate mirror symmetry," confirming the results of experiments led by Wu at Columbia University the month before.

(Symmetry Magazine, Vol. 4, Issue no. 3, (April 2007)

Lederman's experiment was clearly much easier to do (because of the availability of the muon beam) than Madam Wu's, and had a much stronger effect. Two weeks later Valentine Telegdi and Jerome Friedman in University of Chicago would confirm these results in the pion–muon–electron decay chains. While the paper of Lee and Yang won the Nobel Prize, Wu and Lederman did not. (Lederman would win a Nobel Prize for another discovery.) But Madam Wu's own words are a good commentary of the motivation of women in physics.

There is only one thing worse than coming home from the lab to a sink full of dirty dishes, and that is not going to the lab at all.

(*Cosmic Radiations: From Astronomy to Particle Physics*, Giorgio Giacomelli, Maurizio Spurio and Jamal Eddine Derkaoui, NATO Science Series, Vol. 42, (2001), p. 344, Kluwer Academic Publishers, Boston)

C, P Violation and C, P, T Violation

Once the notion of parity being conserved fell, physicists started to look at other symmetries. One such was the symmetry between particles and their antiparticles. Lev Landau proposed that in transforming a particle to an antiparticle, the product of C and P symmetry might be preserved (that is the particle and antiparticle would have opposite parity, meaning that the antiparticle would be a reflected image of the particle). Most of the interactions and transformations indeed conserved CP.

The experiment that showed that even CP was violated also involved Kaons, but this time different varieties of neutral Kaons. Kaons are complicated particles composed of quarks and antiquarks. Two types of neutral Kaons were usually found – one that decayed into two neutral pions with a short decay time, and another that decayed into three neutral pions with a long decay time. (Its antiparticle behaves the same with change in the sign of coordinates.) However, the first has a CP state of -1 and the other $+1$. In 1964, James Cronin, Val Fitch and others did an experiment on a 57 ft long beam tube at the Alternating Gradient Synchrotron in BNL. The fact that the beam tube was long meant that even accounting for relativistic time dilation, the short lived Kaon could not travel the length of the tube. So at the end of the tube, one should not observe two pion decays and only three pion decays from the long-lived Kaon would be observed. Instead, they still observed a few two pion decays (Fig. 10.4). This proved that the long-lived Kaon also occasionally decayed into two pions, which was a violation of CP symmetry. What it means in nature is that certain processes happen at different rates than their "mirror" images, where the definition of "mirror" is slightly different and includes reversing of the charge. For example, a positively charged pion decaying to a positively charged muon and a muon neutrino has the mirror process of a negatively charged pion decaying to a negatively charged muon and a muon antineutrino, but are not exactly equivalent. CP violation is predicted by current theories and observed from the fact that the rate for these two charge conjugated, mirror decay processes is slightly different! This asymmetry has deep importance to why the Universe has abundance of matter over antimatter.

The next step is to check if the C, P and Time (T) are conserved together. That is to check if everything remains the same when with all matter is replaced by antimatter (corresponding to a charge conjugation), all objects having their positions are reflected in a mirror (Parity or chirality change) and all momenta reversed

Fig. 10.4 "Amigo, the boss asked us to check the tube to see if there are any two pions coming out together?" "No, tonto, Go to sleep! CP is never violated. Kaon only decays into three pions"

(corresponding to time inversion). *CPT* symmetry is recognized to be a fundamental property of physical laws. But, like *CP* violation which was also considered impossible, this too may be up for grabs. An example of *CPT* conservation that may be tested is the decay of neutrons into a proton, an electron and a neutrino. One could investigate how an anti-neutron would decay. If *CPT* conservation holds, the flow of changes would be as shown in Fig. 10.2. Indeed, all indications are that *CPT* is conserved together (*CP* is also observed to be conserved in neutron decays).

There are other quantities that are conserved in specific symmetry situations. For example, flavor quantum numbers – baryon number (1 for baryon like proton, 1/3 for a quark, −1/3 for antiquark), lepton number (1 for leptons like electron, −1 for positron), strangeness, charm, bottomness (also called beauty) and topness, color quantum number, isospin, etc. [Baryons are particles composed of quarks and leptons (muons, electrons) are fundamental particles themselves; see later in the chapter.] Each of these comes into play in specific interaction. For example, in all interactions of elementary particles, baryon number and lepton numbers are conserved. For example, a muon can only decay into electron, since both are leptons. An interesting demonstration of the conservation of lepton and baryon number is the decay of neutron into a proton, an electron and an antineutrino, shown in Fig. 10.2. The decay satisfies the baryon number conservation since both neutron and proton have a baryon number of 1, the fact that the electron has a lepton number of 1 and antineutrino has a lepton number of −1, conserves the initial lepton number of 0. Then there is the property of color for quarks which is also a symmetry parameter (section below describes quarks and neutrinos). It should be noted that the names charm, beauty color, etc, do not have the usual meaning or connotation and are just creatively chosen labels for these quantum numbers.

Gauge Theory and Symmetry

The concept of fields exists from Newton's days, when action at a distance was proposed. In physics nomenclature, field is a region of influence, but also an observable quantity like electric or magnetic field, observable because the force on a charged particle is proportional to the strength of the electric or magnetic field. However, one might assign an inherent property at every point in space and time with respect to an electric field – an electric potential Φ. The field and then the force on a particle can then be derived once the potentials are known. For an electrostatic field, all the observable depend only on the gradient of the potential and therefore, a constant potential Φ_0 can be added to the potentials at every point, and the observables (the derivatives of potentials) would remain the same. Lifting the whole electric field region in potential would not be felt by the particles within that region. It is like the distance between two objects. We can change the starting point to go to the railway station and then to the post office, but the distance between the railway station and post office will not change. The addition of a constant potential in the case of an electrostatic field is a specific case of gauge transformation. A similar transformation also applies to magnetic field. Since, the magnetic field satisfies the Maxwell's equation or Gauss's law (Chap. 3),

$$\nabla.B = 0 \tag{10.1}$$

The magnetic field can then be described by the curl of the vector potential A,

$$B = \nabla x A \tag{10.2}$$

But since the curl of a gradient is zero, one can transform A by adding the gradient of a scalar Ω or

$$A \rightarrow A^{\Omega} - \nabla\Omega$$

$$\Phi \rightarrow \Phi_0 + \Phi \tag{10.3}$$

and B would remain unaffected. For electromagnetic fields, a scalar and a vector potential provide the basis and can be changed through this transformation. But, in coming up with a gauge transformation, it is not sufficient to see that only fields be unaffected by the transformation, we also have to ensure that the equation of motion of the particle remains unaffected by the transformation. The gauge invariance, pertaining to the force experienced by a charged particle, should target the equation of motion. The case of a charge in an electric and magnetic field is examined below.

The force in the x direction, on a particle, with charge q and mass m and traveling with a velocity v_y in the y direction, due to an electric field E_x in the x direction and a magnetic field B_z in the z direction (x, y, z are Cartesian coordinates) is given by,

$$m\frac{d^2x}{\partial t^2} = qE_x + qv_yB_z \tag{10.4}$$

As we see that for a time-independent system, this equation is satisfied when we apply the transformations in (10.2). But, when we apply it to time-dependent situation, for example, the case of a changing magnetic field, this transformation becomes insufficient and changes the observables (additional induced electric field exists). Therefore, one has to make the additional transformation on the electric (scalar) potential

$$\Phi^{\Omega} \rightarrow \Phi - \frac{\partial\Omega}{\partial t} \tag{10.5}$$

where the $\frac{\partial\Omega}{\partial t}$ is the partial derivative of Ω with respect to time (only). So the Gauge for electromagnetic fields is given by the transformations given by (10.3) and (10.5). With these transformations and potentials, one can describe the general case of a charged particle traveling in static or time varying electric magnetic fields. The nearest analogy to this is that a body, falling into a uniform liquid, would experience resistive force proportional to the velocity, liquid density, etc. But if the liquid density or viscosity increases with depth, then the body will experience more resistance. If we try to write a single "buoyancy potential" for all situations of this kind, it would correspond to a Gauge. In a quantum mechanical application of this, a wavefunction ψ can be advanced in phase by an angle ϕ by the transformation $\psi = e^{i\phi} \psi$, and the probability of finding the particle in a given location $|\psi|^2$ will remain same, but it will experience different forces at different times. In order to make the forces also invariant to obtain true phase symmetry, we need the constraint,

$$\varphi = \frac{e}{\hbar}\Omega$$

Now, in addition to these transformations, one may lay down specific prescriptions called "Gauge Fixing" such as fixing the scalar potential at one time or location as zero or setting the Vector potential itself as a curl of another quantity and so on. Beyond this point, gauge theories become fearsomely difficult to explain in simple terms and one has to delve into serious mathematics. Furthermore, when physics moved on from classical electromagnetism into quantum mechanical behavior, relativistic quantum field (Gauge) theories (first applied to quantum electrodynamics – QED), were derived and these indeed were the source of new physics that we see since 1950s.

But why are these Gauge theories needed? Basically, this type of gauge transformation prepares the general mathematical analysis for a broad range of situations of forces and particles. What made the gauge theories crucial and sufficient for particle physics is the fact that the gauge theories, which find transformations that constrain the equations of motion, reveal the hidden symmetry in the forces and particle interactions (gauge symmetry). In unifying electromagnetic forces and the weak force that causes the nuclei to decay by beta emission, one needed to transform all governing potentials such that these two phenomena could be described by the same equations of motion. (The potentials are the more fundamental properties of fields and have physicality and are not just a mathematical convenience. This was demonstrated by David Bohm and Yakir Aharanov in a surprising experiment.) It is almost like divining the actual potentials and the patterns in the potentials "seen" by a particle when present in a given force field. The use of invariants associated with symmetries and revealed by transformations becomes the tool for analysis. Feynman and others developed the Quantum Electro Dynamics (QED), a gauge theory that brought Maxwell's classical electromagnetic theory in line with quantum mechanical descriptions. While Herman Weyl is the father of gauge theory, in 1954, C.N. Yang (of the Lee and Yang paper on parity) and Robert Mills defined the gauge theory for the strong interaction. The gauge theories incorporating quantum mechanical treatments have become integral to the present-day understanding of particle physics.

So Gauge transformations are ones that a theoretician makes to examine the symmetry of a particular physical law that governs field and particle, like a connoisseur of objects of art examines by rotating and feeling the piece. Such gauge transformations are, often, beautiful and elegant and may even seem miraculous. Specifically, in arriving at solutions to equations, frequently infinities are encountered and results become indeterminate. Gauge transformations, such as the ones shown above, remove these infinities, while preserving observable properties.

When an unexplained force and associated particles are to be dealt with, a new gauge theory with a new gauge field may need to be created. These theories start from known equations, apply (gauge) transformations to the equations that represent physical laws. If these transformations constrain and apply the equations in such a way that a new set of observable properties are exhibited for say a new situation of fields and particles, without affecting the already known set, then there is a candidate theory for the new situation. The theory would typically "break" a given symmetry for (under) these new situations or impose a new symmetry and introduce new properties and give rise to theoretical concepts.

In a reversed application of these methods, gauge theories are developed which simultaneously explain the behavior of more than one force and one set of particles affected by that force. The more the type of forces that are included in a gauge theory, the greater is the symmetry discovered and greater is the unification. Then one is in the domain of higher symmetries that is described in Fig. 10.5. This leads to what is called the Unification of Forces and Unified Theories. Steven Weinberg

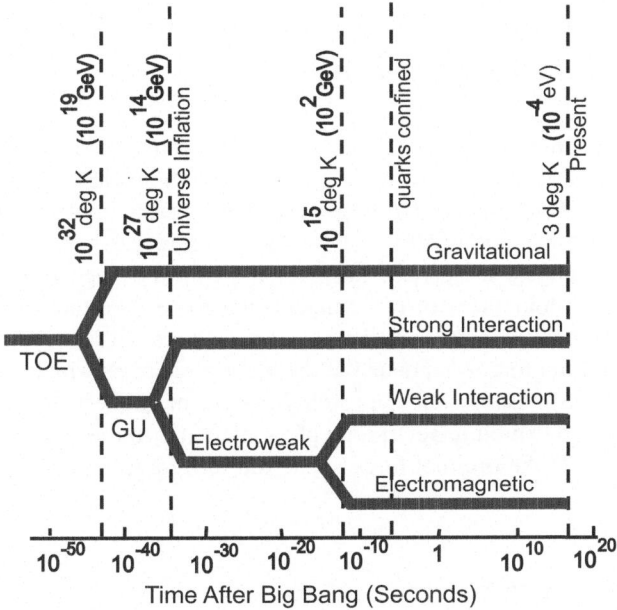

Fig. 10.5 Symmetry breaking and manifestation of different forces at different associated energies of interaction; times when some of the various manifestations appeared are also shown

and Sheldon Glasgow developed a gauge theory that described both the electro-magnetic fields and weak interaction fields and thus came the unification of the two forces. In 1970s, the color quantum number associated with a color field (proposed in 1965), gave rise to the gauge theory of Quantum Chromo Dynamics (QCD). These theories led to the classifications of particles into baryons and leptons. Unification of the four known forces of nature requires the development of a single gauge theory that describes all these forces. The various forces are described by various gauge theories. The designation of the electromagnetic gauge group is U(1). The weak nuclear force and electromagnetism were unified by the gauge theory, proposed by Steven Weinberg, Abdus Salaam and Sheldon Glasgow, for which they won the Nobel Prize in 1979. This combined theory is designated by the SU(2) × U(1) group. The Yang-Mills SU(3) describes the Strong Interaction and the Grand Unified Theory would be the SU(3) × SU(2) × U(1) group. The mani-festation of the different types of forces then comes about by breaking of symmetries.

The physics of particles and indeed fundamental physics is so intimately embed-ded in the mathematical descriptions of symmetries, transformation and gauge theories that even conceptual basis of physics theories are now stated in mathemat-ical terms. In some sense, particle physics of today has become far more inaccessi-ble to lay people than before and descriptions through analogy made more difficult. This is in contrast to Einstein's famous quote "It should be possible to explain the laws of physics to a barmaid".

Forces and Symmetry Breaking

As we saw there are four forces in nature, the electromagnetic force through which electrons, positrons and photons interact, weak interaction which causes radioactive beta decay, the strong interaction experienced by quarks (and therefore protons, neutrons and mesons), and finally the Gravitational forces experienced by particles with mass. Most physicists suspect (and even hope for the elegance) that these four forces all are just four faces of one fundamental force, like the four faces of the Hindu God Brahma, the creator of the physical Universe. It is recognized that this unification of all the forces happened at the highest energies which existed around the time of the Big Bang, and represents a state of highest symmetry. As the Universe cooled to exhibit lower energy phenomenon, symmetries broke spontane-ously to "manifest" the original force as the four different forces (Fig. 10.6).

The Weak Nuclear Force

The forces associated with the SU(N) gauge group is carried by $N^2 - 1$ "boson" carriers (Boson or fermion refer to the statistical properties that the particles have.

Fig. 10.6 Analogy for forces
separately manifesting. The
cream and the milk are the
same and together when it is
hot and boiling. The cream
will separate from the milk
when it cools down

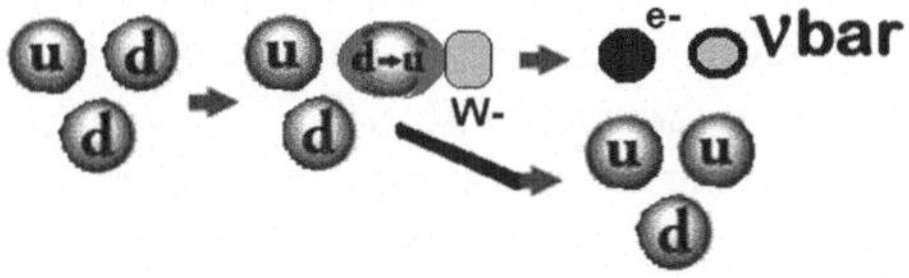

Fig. 10.7 Intermediate stage
of neutron decay into a proton
& W− which then decays into
electron and anti-neutrino

Specifically, bosons have an inherent spin quantum number of 1 and fermions, 1/2.)
So the "weak" force ($N = 2$) would be carried by three Bosons. In the radioactive
decay that Madam Wu observed, for example, the cobalt nucleus transmutates to
nickel with the emission of a proton, an electron and an electron antineutrino. This
is the result of one of the 33 neutrons decaying into a proton and two other particles
(see Fig. 10.5). As we shall see the nucleons – protons and neutrons – are comprised
of so-called "quarks," with properties such as "up"ness or "down"ness. A neutron
has two down and one up quark. In this decay, a down quark in the neutron is
converted to an up quark to give a proton (which has two up quarks and a down
quark). In doing so, it emits a particle W⁻, which is one of the force carriers for
weak interactions. W⁻ then decays into an electron and antineutrino (Fig. 10.7).
W + is involved in the decay of a pion. The Z boson is involved in decays that
change only the spin of the particle.

The quandary was that, while the electroweak SU(2) symmetry explains the near
equal mass of the proton and neutron, it gives force carrier bosons which are
massless just the same as photons. On the other hand, nuclear decay force must
exist within the range of a nuclear size and this is possible only if the gauge bosons
associated with the weak interactions and nuclear radioactive decay had substantial
mass (see Chap. 6). The energy balance in the decay processes and the associated
"neutral currents" (which were measured in CERN before the discovery of W, Z
bosons) predicted a mass of about 80 GeV/c^2 for W and about 91 GeV/c^2 for
another associated Z−, very massive. So there was a contradiction.

The understanding of the weak nuclear force was of crucial importance to the
development of physics, because Quantum Electrodynamic (Gauge) theory, which
had some inconsistencies, was unified with the theory of weak interaction. That is, a

consistent theory which explained the weak interaction and electromagnetism simultaneously was developed.

Spontaneous Symmetry Breaking

In order to reconcile this issue of mass of the W and Z, a theory for spontaneous symmetry breaking (SSB) was invoked. This theory arose from the suggestion by Philip Warren Anderson that the theory of superconductivity may have applications in particle physics. Yoichiro Nambu had started his work in condensed matter physics and was steeped in "BCS" theory that explains superconductivity and the Ginsberg–Landau theory which explains the macroscopic properties of superconductors based on thermodynamic principles. As he stated in his Nobel Lecture, *"seeing similarities is a natural and very useful trait of the human mind"*. In arriving at an unorthodox understanding of the superconducting phenomena through the electromagnetic gauge theory, he surmised that the change to a superconducting state must involve a breaking of the symmetry that is vested in electromagnetic theory. He then came to the realization that this symmetry was broken through the structure of the vacuum state (ground state of the system – a vacuum is not necessarily defined as absence of matter, but filled with quantum fluctuations, consisting fleeting generation and annihilation of particles, antiparticles and electromagnetic waves). Vacuum did not have to abide by the symmetry imposed by the gauge theory. While, in the nonvacuum states (states in which particles interact with fields and forces) the particles are constrained by the gauge. The vacuum ground state has all kinds of possibilities and has many "intrinsic" degrees of freedom. However, Nambu showed that even in this "born free" state, the vacuum state was a charged state, not neutral and this broke the symmetry of the gauge theory and created the superconducting state. An illustration of the theory is as follows:

Let us imagine a marble sitting on top of a conical Mexican hat. The particle may have a tiny amount of freedom to allow a small perturbation and not fall over (the Mexican hat is the shape of the potential hill and well in many of these cases). If someone thumps the table strongly enough, the pebble will fall off the top of the hat and can end up on any side of the rim of the hat. There are infinite possibilities and therefore there is pervasive symmetry. The trough (bottom of the hat) is in contact with table and experiences the vibration of the thump and goes into all kinds of modes, all statistically probable. This is the U(1) gauge symmetry that is applied to the electromagnetism and the electroweak unification brought this type of potentials to the Weak interactions. So, for the Weak Interaction there must be another parameter (ordering parameter) which brings the particle to the specific "vacuum" state, in which it acquires mass. An example of this symmetry breaking is also seen in Ferromagnetic materials, where the atoms, which are like dipoles, are oriented every which way. But the material possesses an intrinsic magnetization and an order parameter such as a magnetic field aligns the atoms of the materials, breaking the symmetry.

He boldly applied the same principle to particle physics and proposed a theory of SSB for particle physics, in 1961. In applying the SSB to electro-weak interaction gauge symmetry, he showed that the quantum fluctuations in the ground state (condensate) break the symmetry. While the bosons would acquire mass in some of the broken symmetries of SSB, some had no mass. What the order parameter for Weak Interaction is and how this leads to the massive W and Z boson and indeed mass of any particle is the crux of modern high energy physics.

The Journey to the Standard Model

Inspired by the work of Nambu, three papers appeared in the Physical Review Letters in 1964, one each by Robert Brout and Francois Englert, by Peter Higgs and by Gerald Guralnik, C. R. Hagen, and Tom Kibble. These proposed a gauge theory, in which a pervasive specific scalar field would be added to the known set of fields. A vector field like electric field is undesirable because it would select a preferred direction and violate spatial homogeneity. For example, in the case of the Mexican hat, a fan blowing in a specific direction would break the symmetry, but that would not be spontaneous and would preclude all the other permitted directions of fall. When a gauge field (electromagnetic or Strong) associated with a particle interacts with this scalar field, gauge symmetry spontaneously breaks and creates massive particles. This is a bit like a baseball or a cricket ball pitched/bowled with a spin. If there were no air, the ball's motion is unaffected in its path except for its spin and motion in the thrown direction. But in the presence of all pervasive static air, the air and the spin of the ball interact and differential air pressure is created to push the ball one way or another or change speed, break symmetry and give rise to a particular behavior.

This scalar field has now come to be known as the "Higgs Field" and the force exchange particle, the "Higgs boson" with zero spin. Peter Higgs himself credits Anderson and Nambu for the insights and originality of ideas but added that their work was not understood until he wrote specific passages in his paper. The Higgs field is now postulated to pervade the Universe. However, this field is not like ether, which was originally proposed as the frame of reference. The Higgs field is inertia generating medium, in which particles like photons can slip through without friction, but others experience resistance to varying degrees. So, once unification of electromagnetic field with weak interaction was proposed in 1960 by Sheldon Glasgow, in 1967, Steven Weinberg and Abdus Salaam invoked the Higgs mechanism to make the electroweak theory consistent with massive W and Z bosons. With this, the Higgs field has now become the cornerstone of the model of all particles and fields, "The Standard Model" (SM), which presents the organization of fundamental particles.

While the Standard Model is a prize any one would be proud of, the journey to it was its own reward. The proposal of quarks, discovery of tau lepton and discovery of neutrinos, all the antiparticles and the fabrication each brick of the edifice of the Standard Model theory, all are stories that could be told in million pages. The beginning of this journey could not even be traced to a particular discovery, because the SM incorporated all the knowledge on particle physics, gained over two centuries and made them all consistent. But it is clear that the lead stars in the Standard Model are the quarks, because they are so "colorful."

Quark, Quark: No Ducking the Issue

It would seem a long time back, but Yukawa seemed to have come up with the description of the nuclear force. In 1961, Murray Gell-Mann proposed the "Eightfold Way," organizing the zoo of mesons (spin = 1, spin = 0) and baryons (spin = 1/2), according to octagons (one for each spin) with the vortices corresponding strangeness and electric charge. There was an octagon for each spin (Fig. 10.8).

Similar octets, nonets, and decuplets are given for other mesons. This quelled the riots for a while, but there was considerable distress over the number of particles. Murray Gell-Mann and George Zweig independently proposed that all these particles could be constructed by assuming that these were constructed of subparticles, Gell-Mann called them quarks first (just to give it a funny name) but then found that it would match with James Joyce's invention of the word in "Three quarks for muster Mark" in his novel Finnegan's Wake. Zweig named it aces, but lost out in the name game, possibly because he went on to become a biologist. It was a matter of simple arithmetic of permutation and combination to come up with the properties of quarks. What charges and how many quarks do we need to build the non-strange particles (such as protons, neutrons and delta with isospin = 1/2) and how many to build integer spin particles, what different values of charge must these quarks have, etc. It turns out that if the quarks (antiquarks) could have an electric charge of $\pm 2/3$ or $\pm 1/3$ and each was a fermion itself (spin = $\pm 1/2$), one would get what one wants. Specifically if the up quark (u) has 2/3 charge and down quark (d) has $-1/3$ charge, then the nonstrange particles would be made of three quarks

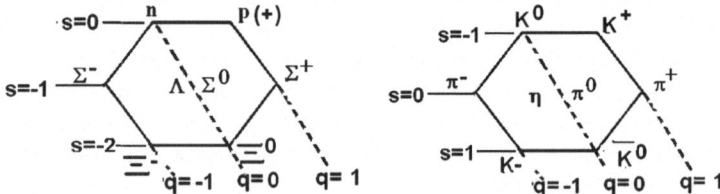

Fig. 10.8 The eightfold way Octet, showing the organization of mesons and nucleons (Gell-Mann), which, in turn, is explained by a composition of different types of quarks (Gell-Mann and Zweig). On the left, so-called baryons with a spin of 1/2. On the right, a similar Octet for mesons (integer spin)

(hence the James Joyce reference). For example, the proton would be made of uud, while neutron would be made of udd. The delta++ would be made of uuu and delta− with ddd, and so on. Integer spin particles should have two quarks and this way, meson structure is also addressed − they would be made of a quark and an antiquark (see below − discovery of J/psi). An example is the pion π+ which has one up quark and one down antiquark, to give a charge of one and zero spin (The quark and antiquark being different types would not annihilate each other, also see below.) Each quark also has a baryon number of 1/3. But then there remained the strange particles which did not conform to this group. In order to explain these he postulated the strange quark.

For a while Gell-Mann thought that the idea of quarks was just a mathematical construct. No wonder, because, no particles with fractional charge had been found and there was no trace of a particle called quark. Even today, this is an issue. Quarks cannot be seen by themselves. Quarks are free within a composite particle, but if one tries to pull them apart, the strong interactive force increases and stays at a constant force of 10,000 N (humongous force for such a tiny particle. It is like an apple being pulled in by a force of 10^{32} N!).

Color Quantum Number and Asymptotic Freedom

There was an unhappy problem, a principle called Pauli's exclusion principle, which forbids particles with identical quantum numbers (such as uuu, ddd and even uud, etc.) to coexist in the same state. This argument also supported the belief that Gell-Mann's description was only mathematical and could not happen in reality. Boris Struminsky first broke the ice stating that an additional quantum number would be adequate to avert the exclusion principle, because each quark could then be different. In 1965, Y. Nambu, M.Y. Han and O. Greenberg proposed the SU(3) gauge theory with this extra quantum number, now called "Color." So, according to this now accepted theory of quantum chromo dynamics (QCD), which parallels the QED, quarks come in three color charges, red, green and blue. The magnitude, the addition of color charges is not straight forward and QCD deviates from QED in this regard. The color charges would feel forces in a potential similar to electromagnetic potentials.

James Bjorken proposed that if the protons had more fundamental point-like particles, these would be found in the so-called deep inelastic scattering experiments. The experiments in SLAC showed spectacularly that protons did contain point-like parts, which Richard Feynman, in spite of his belief that these were the quarks, called "partons," exhibiting the reluctance of the physics community in identifying them as quarks. Paradoxically, the experiments found that these particles were rattling around inside the proton quite happily as if they were free. In 1973, following up on the suggestion of Y. Nambu, M.Y. Han, O. Greenberg, the strong interactions were concluded to be mediated by gauge bosons, the gluon. David Gross and Frank Wilczek, working at Princeton, and David Politzer, working

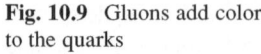

Fig. 10.9 Gluons add color
to the quarks

independently at Harvard, proposed that the quarks were imbedded in a pool of these gluons. Unlike the colorless cousin photons, these gluons themselves have color (mixture of two colors). Quarks, which on their own have little color, are swamped by the colored gluons and acquire more color (Fig. 10.9). The farther one goes, the pool of gluons is thicker and the quark appears more and more colorful. The gluons carry color and anticolor between quarks, constantly changing their color. Since this color change represents the interacting force between quarks, the force also increases with distance. Conversely, the attraction between quarks grows weaker as they approach each other (Asymptotic Freedom). It took 30 years, but the trio did get their Nobel Prize for this (Fig. 10.9).

Charm is the Brother of Strange

It had been observed for long that certain massive particles "strangely" decayed through weak interactions and had a longer life than expected for such massive particles. But these were expected to obey laws of strong force and react rapidly in a medium. These were also "strangely" produced in pairs (strange particle and strange antiparticle). So these particles were called "Strange." Now we know that strange particles behave strangely because they consist of the so-called "strange" quark, an unfortunate, vicious circle of names, because there is nothing strange about this quark, it is simply a different flavor of a quark. The strange quark could decay into an ordinary up quark, an electron and a neutrino and therefore could change flavor. The strangeness came from the fact that associated with all particles is a strangeness quantum number and a strange particle has no smaller mass particle with the same strangeness quantum number. So it is a metastable particle with relatively long life time, before it decays through weak interaction without conserving strangeness. An example of this is the lambda particle decay, which has an up, a down and a strange quark and decays into a proton or a neutron and a pion, which do not have a strange quark. While they are other different decay possibilities for a strange particle, one mode was never observed. Murray Gell-Mann and Sheldon Glasgow noted that the electroweak theory predicted "strangeness changing neutral currents" (SCNC), a term, Sheldon calls "incomprehensible exemplar of gibberish" and also "Ecch! An unwanted, unseen and despised ... effect" (Interactions by Sheldon Glasgow, Warner Books, 1988). Glashow (aided by John Iliopoulos and Luciano Maiani) then reasoned that if the strange quark had a partner, the "Charmed Quark" (now called charm quark), the problem goes away because the generation of SCNC is cancelled by participation of up, down, strange and charm quarks. True to its name "charm" (as in charm bracelet), the "charm quark" averted the evil of SCNC.

As was stated before, strange particles are always produced in pairs as particles and antiparticles. These would not annihilate each other immediately and a bound state of these could exist, bound state like a hydrogen atom of proton and electron (for example, the positronium, bound state of positron and electron, also exists). Therefore, now the final proof of the electroweak theory hung in the discovery of the bound state of the charmed particle and charmed antiparticle. In the early 1970s, this would be the holy grail of particle physics. In a spectacular and unbelievable coincidence, two groups announced the discovery of two particles on November 11, 1974. The team led by Samuel Ting, at the first ever and large alternate gradient synchrotron AGS in Brookehaven (BNL), announced the discovery of the J meson and the team headed by Burton Richter at the first ever high energy linear accelerator at the Stanford Linear Accelerator Center team, announced the discovery of the Psi particle. The two teams did not know of each other's discovery. Instead of being called the "November Miracle," it is now called "November Revolution," presumably because physicists do not believe in miracles. It turns out that these two particles are same and are the "charmonium," bound state of charm quark and charm antiquark. The discovery of this meson promptly earned the two physicists, the Nobel Prize, in 1976.

The charm not only addressed the SCNC, it also established a pattern for the quark set. Now there was an up quark and a down quark, bolts and nuts of the matter around us, the strange and the charm quark that arrive in particle packages from the cosmos. In 1977, a fifth baby, the bottom quark, arrived unexpectedly, discovered by the Fermilab group led by Leon Lederman's team (Yes, he did get the Nobel Prize for it). This discovery was ahead of the theory. All eyes then were looking at the delivery rooms of the particle detectors for signs of the twin of the bottom. It would take 18 years and the top quark was indeed discovered.

Neutrino, a Wisp of a Particle

One other group of constituents, very important but not well covered topics so far, are the neutrinos (and antineutrinos). The neutrino was proposed by Wolfgang Pauli to account for the angular momentum conservation in neutron described above. He introduced this particle to the conference on Radioactivity by addressing the audience as "Dear Radioactive Ladies and Gentleman." In 1956, Clyde Cowan, Frederick

Reines, F. B. Harrison, H. W. Kruse, and A. D. McGuire discovered the neutrino in a beta decay experiment where the signature was the gamma rays emitted in the final step of a neutrino interaction with a proton. It took a long time of about 40 years for them to receive a Nobel Prize for this. Later more types of neutrinos were discovered. Neutrino physics is very important for astrophysics and solar physics and discrepancies in solar emission of neutrinos are still a problem in the Standard Solar Model. Neutrinos interact very little with matter and their tiny mass remains unmeasured. But some things are known about them: they interact with matter only through weak interactions (W and Z bosons) and they are left-handed (intrinsic parity) – meaning that their spin is always in the direction of the curled fingers on the left hand, when the velocity is pointing along the thumb. Their role in cosmology, dark matter, etc., too is a hot topic. If neutrino had more than 50 eV/c^2 mass, the Universe would not have expanded. The Standard Model assumes that neutrinos are massless. The first indication that neutrinos have mass was obtained in an experiment in Italy's Gran Sasso Laboratory and from Super-Kamiokande experiment in Japan.

The Standard Model

For long, the atom and the nucleus could be treated on different basis. Quantum physics of electrons and of light could be treated with quantum electrodynamics and the nuclear interaction involving quarks with strong interaction. But the weak interaction involves all these particles and so they needed to be brought to same consistent theory. The Standard Model is developed in order to make the quantum electroweak theory and quantum chromo dynamics (QCD), internally consistent with each other, as a quantum field theory. As a result of this theory, the bewildering array of particles does not exist any more. Without much ado, the conclusions of the theory are summarized below, partly because the theory is much too complex.

The essential prediction of the Standard Model is that there are three generations of two types of matter particles – hadrons (quarks) and leptons (both fermions with spin quantum number = 1/2) and one type of force particles (bosons with integer spin). These are considered to be fundamental particles that make up the matter and field interactions (see Tables 10.1 and 10.2). There is an antiparticle corresponding to each of the particles. All the protons, neutrons, and mesons are made of quarks and/or antiquarks. The force mediating particles, the photon, the W (+ and −), the Z and the gluons also are part of the standard model.

Table 10.1 Matter Particles in Standard Model, quarks are hadrons and the other two rows are for leptons. All these are fermions with a spin of 1/2. There are corresponding antiparticles

First generation	Second generation	Third generation
Up Quark	Charm Quark	Top Quark
Down Quark	Strange Quark	Bottom Quark
Electron	Muon	Tau
Electron Neutrino	Muon Neutrino	Tau Neutrino

Table 10.2 Force mediating particles. These have a spin of 1

Electromagnetism	Weak interactions	Strong interactions
Photon	W+,W−, Z0 bosons	8 Gluons

Each of these particles has now been discovered (The tau was discovered by a SLAC – Lawrence Berkeley Lab team, headed by Martin Perl and the top quark was discovered in Fermi National Laboratory). The gluons were observed at the DESY laboratory experiments in PETRA and PLUTO (DORIS) in 1979. This seemingly simple classification hides behind it, considerable amount of physics and mathematics, as the foregoing sections indicated. This is, in essence, the simple but elegant façade of the edifice of particle physics which was built brick by brick through carefully formulated theory and from diligent observations on terrestrial radioactivity, cosmic rays and last but not least the collisions of accelerator beams. This edifice, SU(3) × SU(2) × U(1), has the QCD sector and the Electroweak sector. In addition to the symmetries, incorporated in the above sectors, one finds other "accidental" conservation symmetries, such as total number of electrons in the universe and also the traditional Poincare symmetries such as conservation of momentum, energy, etc. The Standard Model has an important sector, namely the Higgs Sector and the associated Higgs Boson (spin = 0), which remains undiscovered. One of the ardent quests of present-day accelerators is the discovery of this Higgs boson, called the God Particle, because it bestows solidity and mass to all other particles. It is believed that the mass of the Higgs boson can be found by an accurate measurement of the W or Z boson. Taking a global fit of all the data from various experiments, the upper limit for the mass of the Higgs particle is 150 GeV/ c^2 (Klaus Monig, Physics 3, 14 (2010)), while some others even give a lower limit of 114 GeV/c^2.

The standard Model has three generations. Why are there 3? Are there more? If yes, what are they? If not, why not. This is like the question of "is there extraterrestrial life?". The Standard Model is a model for Unification of particles. While, the electroweak and the Strong interaction have been brought into a dating situation in the Standard Model, they are not married. What may we find that will unify the three forces to obtain a Grand Unified Theory? There are a few candidate models, including a theory called Supersymmetry which postulates companion superparticles (or sparticles), but so far any evidence in favor of Supersymmetry has eluded us. The SM has zero mass for the neutrino, but the observed "neutrino-oscillations" in which the neutrino changes its type, implies some neutrino mass. The Standard Model also includes some expectations for the Cosmological parameters. It has a prediction for the Cold Dark Matter and for the dark energy. This prediction has too small (4%) a dark energy/matter which, according to Standard Cosmological Model, should be over 90%. Much is expected of the future accelerators and detectors in solving even more difficult questions from the physicist's insatiable minds.

Chapter 11
Collision Course

When boys play with toy trains and imagine spectacular collisions, they first dash them into walls. When they get other boys to bring their trains, they dash their trains into those others, to imagine even more spectacular collisions. Physicists know the benefit of these collisions and have learnt to exploit them. The way to think of a high-energy particle collision is that when an energetic particle hits another oncoming high-energy particle or a target (a solid, liquid, or gas consisting of other particles with very small energies), the identity of the colliding particles is lost and a soup of matter and energy is created. Then, new particles are born from this soup with various probabilities, while conserving certain basic parameters that went into the collisions. In earlier accelerators, the particles were only smashed into stationary targets and the accuracy of the beam targeting was not an issue. But, in such a scheme, the incident particle and the new particles carry away considerable energy after the collision, without using all the incident energy to create the new particles. This is similar to a ball thrown against a wall; the ball does not leave much energy behind at the wall as it bounces back. This is because the net momentum of the particle has to be conserved and the incident particle and any new particle, created in the collision, have to carry this momentum. The energy available for the new reaction or new particle creation is then only what is left after supplying the kinetic energy of the outgoing particles.

However if there are two particles (labeled 1 and 2 with velocities v_1 and v_2, masses m_{01} and m_{02} and Lorentz factors, γ_1 and γ_2), then the net initial momentum is $(\gamma_1 m_{01} v_1 + \gamma_2 m_{02} v_2)$. Now, if identical particles, $m_{01} = m_{02}$, are traveling in opposite directions, $(v_1 = -v_2, \gamma_1 = \gamma_2)$, then net momentum is zero. In other words, in conserving total momentum, the outgoing particles do not have to carry away any net momentum and therefore no energy (or may carry arbitrary canceling momenta). Therefore, in principle, all the energy that went into the collision is available. For a head-on collision of two identical particles with mass m_b, velocity v_b and energy E_b, the total energy of $2E_b$ is available for the reaction to create new particles. But for a fixed target collision [target particle (nominally a proton) with mass m_t is stationary], the energy available is only equal to $\sqrt{(2m_t E_b c^2)}$ (approximation when the beam energy is much larger than the rest energy of the beam particle and the stationary target particle – usually a proton in the nucleus). The threshold (minimum) energy required for creating a particle with mass M is then given by, $Mc^2 = \sqrt{(2m_t E_b c^2)}$, or $E_b = (M^2 c^2/2m_t)$ with $m_t = m_p$, mass of the proton in the target. On the other hand, each of the two particles colliding head-on, need to have a minimum energy of only $Mc^2/2$. Therefore the fixed target beam has to have an energy about (M/m_t) greater than colliding beam energies. If an incoming beam is required to create say, the W boson with a rest mass $Mc^2 \sim 80$ GeV, one would require about 40 GeV for each head-on colliding beam, but about 3,200 GeV for the fixed target experiment, since mass of the target proton is about 1 GeV! Therefore, as the particles to be discovered become heavier, fixed target experiments become practically infeasible. Therefore, recent decades have seen great progress in the center of mass energy (sum of the energy of the two colliding particles) (see Fig. 11.1)

This concept of obtaining the above "center-of-mass" collision in order to fully use the particle energies had been considered by the Norwegian engineer and inventor Rolf Wideröe (1902–1992), who, as we saw in the previous chapters, also pioneered many accelerator concepts. He had constructed a 15-MeV betatron in Oslo and had patented this idea in 1943, after considering the kinematic advantage of opposing beam collisions to get larger momentum transfers. The

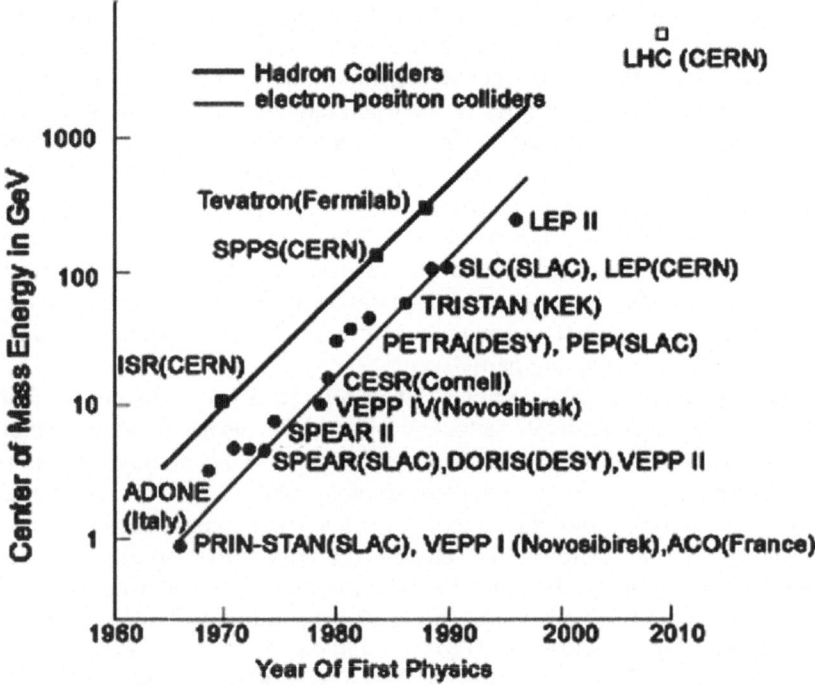

Fig. 11.1 The increase in available center of mass energy in particle collision experiments vs. time (SLAC Beamline online, Spring 1997, p. 36)

Princeton–Stanford group seriously considered exploiting this idea. In 1956, Burton Richter and Gerry O'Neill proposed building two tangential colliding rings. Andrei Mihailovich Budker also started the VEP-1 electron–electron collider in 1961. While these physicists were focused on this kinematic advantage of the scheme, Bruno Touschek in Frascati was thrilled at the potential physics use of the hitherto unrealized amount of energies available in these collisions. He pointed out that in the past, only photons had served as point particles in space for probing the structure of matter, but at high energies, the dynamics of particle interactions and their behavior could be examined also in precise pin-points of time. One way to look at this is that energy and time are complementary parameters – Heisenberg's uncertainty principle states that the product of variation or uncertainty in each, that is $\Delta E \times \Delta t \sim h/(2\pi)$, h being Planck constant $= 6.626 \times 10^{-34}$ J s or 4.1×10^{-15} eV s. So, when the energy – and therefore the spread in energy – is large, the spread in time is small (This is analogous to having a high shutter speed on a camera to get short time exposure). For example, if one has particle energies of 100 GeV, one could explore events occurring in time scales of 10^{-25} s, time scale over which the early big bang universe was being born (One can then see what phenomena happen in this short interval.). Of course, there is always the drive to get high energies so as to have a high probability of "seeing" high-energy particles for discovery and study.

Collider Scheme and Important Parameters

In particle colliders, two beams of particles would be accelerated, first in a synchrotron or a linear accelerator (LINAC) and then made to collide coming from opposite directions. In many cases, separate storage rings would accumulate the particles from an accelerator, so that the colliding beams can have a large number of particles in the collision. With this development, the schema of the apparatus for investigation of high-energy physics is complete – accelerators provide high-energy particles and colliders extract this energy from particles to form new particles and reactions.

For Europeans, the electron–positron collider and the massless photon resulting from the annihilation as pure energy, was very appealing. Bruno Touschek, a brilliant man, taught at a University in Rome and Pisa, but because of Italian rules on promotion, etc., his academic career never really took off. When in Gestapo jail during the war, he turned into a scientist–cartoonist while applying himself to broad topics in physics (Fig. 11.2). Later in 1960, he held a seminar during which he proposed an electron–positron collider, with the two beams circulating in the opposite directions in the same ring. The first such Collider AdA (coinciding with Touschek's favorite aunt's name) was built in 1960 in Europe in Frascati, Italy and proved to be a success. Touschek is much beloved in the scientific community. When he died in 1978, his colleagues' eulogies reveal the man.

Fig. 11.2 Cartoon by Bruno
Touschek, showing scientists
arguing about magnetic fields,
currents and forces. This
perfectly fits into the
stereotype of Italians using
their hands expressively, in
all conversations

MAGNETIC   DISCUSSION

"He was the initiator, but also the element of continuity during the ten golden years of the
Laboratories (of Frascati), the person that had a great idea and allowed it to be materialized
by others; his scientific and human qualities, I believe, were decisive in maintaining the
connections which have been essential in achieving success" (by F. Amman). So Bruno
Touschek will always be remembered as a man who "led an intense and vigorous life and
one who, by his example and friendliness, helped others to achieve greater happiness and
awareness in their own lives" (by P.I. Dee)

Today, the large synchrotron collider, using the electron positron collisions, is
the CERN's 27 km long LEP (Large Electron Positron) collider, with a collision
energy of 92 GeV. This corresponds to an energy of over 2,500 GeV, in fixed target
collisions. The Tevatron in Fermilab, USA, completed in 1985, was till recently the
highest energy (~1 TeV on ~1 TeV) synchrotron proton–antiproton collider.
Stanford's Linear Collider, with the energy of 50 GeV on 50 GeV, is the first
collider of its kind with no bending magnets, and it operates in a long straight beam
line. The largest ever accelerator of this type, the Large Hadron Collider, has just
been commissioned in CERN.

The particle choice is made on the basis of their ease of production, the life time at
low energies (at high velocities, relativistic time dilation increases the life time in the
lab frame). Commonly, electrons, protons and heavier nuclei are used. Antiproton
and positrons beams are also commonly used. The advantage of proton–antiproton
and electron–positron colliders is that they can be accelerated and stored in the same
synchrotron ring beam tube. The electron–positron colliders produce fewer species
of particles because they are both fundamental particles. The proton–antiproton
which comprise many types of partons (quarks) and therefore, the collisions are
somewhat messier creating a larger variety of particles. At collision energies below
2–3 TeV, the primary reaction is the quark–antiquark fusion. However, the accumu-
lation of antiproton has a limit, due to the first step of having to create and collect the
antiprotons. Therefore the beam intensity of antiprotons is low, giving a low event
rate. Proton–proton colliders can have much higher event rates, but below 2 TeV, the
quark–quark collisions create an even larger spectrum of particles which can crowd
the signal of a particle one is looking for (signal to noise ratio becomes poor).

However, at collision energies above 3 TeV, the difference between proton–antiproton collisions and proton–proton collisions decreases substantially, because in both cases the primary process is gluon fusion. The higher beam intensity available in proton–proton colliders makes it more desirable despite the disadvantage of needing an additional ring, since the second high intensity proton beam is more easily created than a lower intensity antiproton beam.

One disadvantage of moving from a fixed target to a 2-beam collider is the fact that the fixed target provides a rich and dense collection of particles, while a beam is like a gas at much less than 1 mm of Hg pressure. Whereas the fixed target collision produced a trillion events in the 30 GeV AGS at Brookhaven, the proton–anitproton collisions at the Fermilab Tevatron has produced only a few million events. Therefore, the key requirement is that the beam intensity in colliders (the number of particles in a beam which is usually sent in bunches called buckets), be as large as possible to maximize the number of interesting collision events and increase the probability of seeing a specific reaction or particle. The frequency with which the particles collide, is determined also by the so-called reaction cross section, which is analogous to the cross sectional area of a target – larger the better. A reaction cross section is related to the probability of a specific reaction to occur. For example, the cross section for ionizing water vapor by a 50 keV proton is about 8×10^{-20} m^2. The rate of collision of an incident particle with target particles is $= n\sigma V$, where n is the particle density in the target, σ is the collision cross section and V is the velocity of the incident particle. Since the collisions are not exact analogies of collisions of objects, the cross section and rate of collisions are, in general, dependent on the energy, angle of incidence, type of reaction one is interested in, and type of particles involved in the collision.

In colliders, the beam is not a continuous one. It is sent out in bunches and therefore, the term beam intensity has to be redefined. The term beam luminosity represents the equivalent characteristic. If one target bunch has N_1 particles, physical cross section A and length L, then the particle density in the bunch is $N_1/(AL)$; the rate of collision of an incident particle with this beam is then $(N_1 \sigma V)/(AL)$. When N_2 particles in n_B bunches of particles are incident on the beam, the number of collisions per second (rate) would be

$$f = (N_1 N_2 n_B \sigma V)/(AL) = (N_1 N_2 n_B \sigma)/(AT), \qquad (11.1)$$

where $T = L/V$ is the time taken by the particle to cross the bunch. With $1/T \sim f$, frequency of particle arrival, the collision rate is given by

$$R = (f N_1 N_2 n_B \sigma)/A = L\sigma, \qquad (11.2)$$

where L is the "luminosity" $= (f \, n_B \, N_1 N_2)/A$.

As can be seen the luminosity is defined to be the rate of interaction for a given cross section and is therefore independent of the type of reaction. Clearly, it is important to bring the beam bunches to a tiny focus so that the density of particles is high (beam cross section A is small). In a collider, the last section following the accelerator and just before the collision point (inside a detector) is called the interaction region, the function of which is to bring the beams into a tiny focal point (see later) and steer them to collision.

An accelerator–collider assembly is mainly characterized by two parameters – the energy of the colliding particles and the beam luminosity. All parts of the assembly are designed to maximize these parameters. As had been seen before, particle energies are limited by power of the magnets and the length of the beam path and by the ability to accelerate. The luminosity is limited by how many particles one can pump into the beam, how fine a beam we can make and how frequently we can make them collide. For synchrotron colliders, there are many conflicting needs. An example: In order to have high-energy beams with a given bending magnetic field, the radius of the ring has to be large. But a large radius (longer path) implies that the frequency of revolution (number of times a bunch arrives per second at the collision point) is small reducing the luminosity. This can be compensated only by having more bunches implying greater source of beam particles. There are many such tradeoffs in a collider design and the final choice depends upon the goals of the experiment.

Multiple Rings and Storage Rings

As we saw above, one of the important needs for a collider would be to collect particles into a high intensity beam and then have them collide. But a synchrotron can only accelerate particles by a certain factor, usually 5–10. This factor is limited due to several reasons, one of which is that the magnitude of the magnetic field has to correspond to the particle energy. If the incoming particles have too small an energy, the magnetic field required to bend them would be too small. Then particles might be lost, because the field is insufficient to confine them. Typically also, the magnetic fields cannot be raised over a large range for a given set of magnets, limiting the maximum energy for each ring. In superconducting rings, at very low fields, the field errors become too large due to, so called, "persistent magnetization currents". Therefore, following the original idea of Matthew Sands (adopted for example in the Fermi Main Ring), a high-energy synchrotron–accelerator complex would consist of stages of (multiple) accelerators. Each successive stage is larger in circumference (because of the limitation of magnetic field strength as the energy increases). The larger ring of a stage would have to be filled from injection of multiple turns from the previous smaller stage by the factor equal to the ratio of their circumference. Once filled, the final ring must store the beam till the experimenter is ready for the collisions (Fig. 11.6). The smallest ring is typically filled from a LINAC connected to a particle (ion or electron or antiparticle) source. Often, in particle–antiparticle (like electron–positron or proton–antiproton) collisions, antiparticles have to be produced and accumulated in a separate ring. An example of the antiparticle is the Fermilab antiproton production scheme is described below.

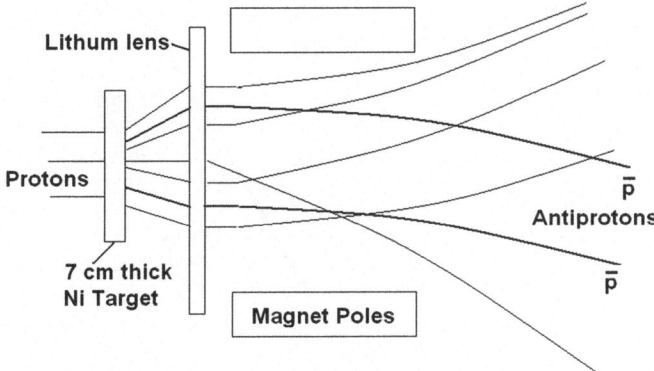

Fig. 11.3 The process for producing the antiprotons from a nickel target

The Target for Generating Antiprotons

A beam bunch of 120 GeV protons from the Main Injector (formerly the Fermi Main Ring) is directed at a nickel target every 1.5 s (Fig. 11.3). One in a million of these hits produces antiprotons (also called pbars). Since these antiprotons are emitted from the target at different angles, they are focused by a Lithium lens. The particles coming out of the lens are sorted out by a magnet bending only the antiprotons into the receiving pipeline; the rest of the unwanted particles are bent outside the aperture of the receiving pipe.

The Debuncher

The debuncher works on the same principle as the longitudinal focusing described before. The antiprotons arrive in bunches because protons that create them come in bunches and within a bunch the particles will have slightly different energies. Since the antiprotons have a large energy, about 8 GeV, their velocities are nearly the same – close to the speed of light in vacuum. Now when their paths are bent by a magnet (see Fig. 11.4), they separate since higher energy particles have larger bending radius and therefore a longer path to travel when they are turned. The antiprotons are turned back by another magnet and arrive at an RF cavity, which is the debuncher. The higher energy particles would arrive later than the particles with lower energy. Now if the accelerating phase of the RF is arranged such that the lower energy particles arriving first see a higher acceleration voltage than the more energetic particles arriving later, then the lower energy particles would be accelerated more and the more energetic particle would be accelerated less or decelerated. This way, the energy spread would be reduced. After several passes all particles would have the same energy. But since, the particle velocities have remained the same, during the period of differential acceleration, the leading

Fig. 11.4 The debuncher
reduces the spread in
antiproton energy while
spreading them in time

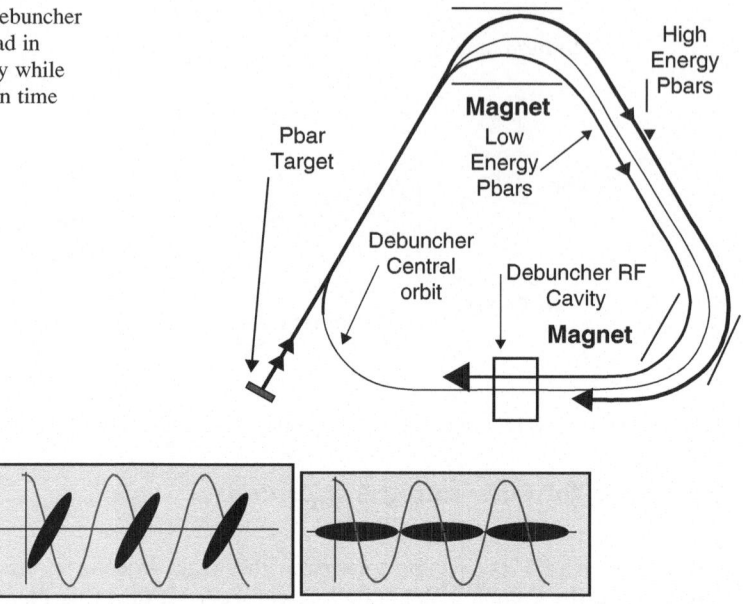

Fig. 11.5 The bunching in energy and spreading in time. Vertical axis is energy and horizontal axis is time. The wave form is a reference for time and the amplitude relates to the dynamic nature of this debunching process, in which some particles would get accelerated more and some less in one cycle and then catch up in the next cycle etc. (Courtesy: Fermi National Accelerator Laboratory, Balvia, IL, USA.)

particles have increased their lead over the lagging particles and therefore the bunches have a longer time spread and are debunched in time (see Fig. 11.5) (This reminds one of the uncertainty principle, doesn't it?). As one can see, the debunched particles see different accelerating voltages depending upon their time of arrival and therefore the process would be dynamic.

Stochastic Cooling/Accumulator

The remaining energy spread is equivalent to the temperature of the beam with randomness in energy and makes the particles wander away from their path. This is reduced by what is called "Stochastic Cooling". In this method, which won Simone Van der Meer, the Nobel Prize, if a particle has a deviation in the path (this is actually not one particle, but a statistical group of particles), then a very small error signal is detected by a sensor, which senses the deviation in terms of the path change. This signal from the thermally stable, highly sensitive sensor is amplified and with a feedback system, a "kicker" delivers a kick to correct the path of the particle. Both the RF and the stochastic cooling are more essential for antiprotons, because they would disappear in annihilation process if their path is not controlled accurately and do not remain away from the tube walls. After going through the

Fig. 11.6 Schematic of the
20 TeV US Superconducting
Super Collider Linac-Linear
Accelerator, LEB, MEB,
HEB – Low-, Medium- and
High-Energy Boosters (SSCL
Site Specific Conceptual
Design Report, July 1990)

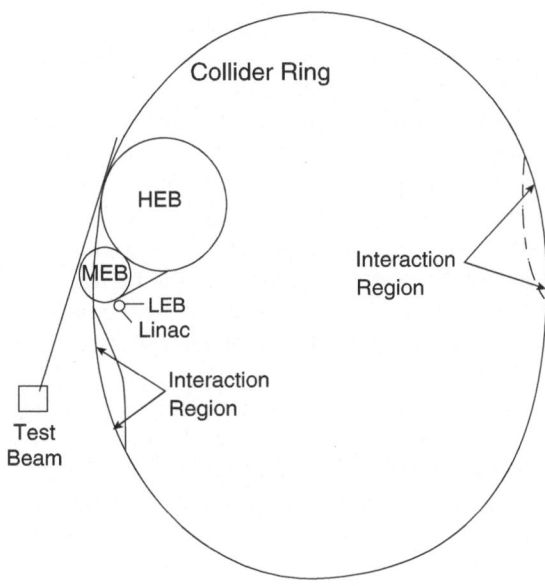

cooling process, the antiprotons are stacked in the "accumulator" and stored for several hours in very high vacuum conditions, until it is time to inject them into the accelerator chain.

The Collider Assembly

Figure 11.6 gives, as an example, the scheme planned for the Superconducting Supercollider (SSC) (This machine was not completed because of cancelation of project.). In this plan, an ion source would generate a low-energy beam of protons; a LINAC would accelerate the protons to 600 MeV; a low-energy booster (LEB) synchrotron would increase the energy to about 11 GeV; a medium energy booster (MEB) to 200 GeV, a 11 km circumference high-energy booster (HEB) super-conducting synchrotron would raise the energy to 2 TeV, and then the beam would be injected into the superconducting Collider with a circumference of 87 km. Similarly, a second beam would be injected in a counter direction into a second ring of the collider, to form colliding proton beams. Then the beams were to be deflected towards each other into a collision, at four intersection points.

The Accelerator Cell

Each of these beams would have to be bent into the design orbit path using dipole magnets with magnetic field oriented in the direction perpendicular to the plane of the ring, guided along any straight section and guided out of the synchrotron to provide test beams. The beams would have to be kept focused using quadrupole magnets that are oriented alternately to provide the alternating gradient focusing described before. In general, the dipole and quadrupole magnets always have error fields in that they are never purely dipole or quadrupole magnets. These error fields can make the particles wander out of the desired focused region (dynamic aperture), or even scatter them out. A large loss of particles this way is clearly not desirable. For beams with high luminosity, even a small fractional loss might not be tolerated, because the escaping particles would hit the beam pipe walls and knock out atoms, releasing gases that were adsorbed by the surface. This would degrade the vacuum, causing a strong discharge of ionized gases that could cause further heating and even destroy the beam pipe.

The error fields can be corrected by opposing higher order (higher number of poles) magnets such as hexapole, octopole, etc. For large synchrotrons, these correction magnets can be located periodically and locally, since the particles do not move significantly out of the desired path due to the errors in fields and hardware errors over a distance of say a FODO cell. It is therefore enough to correct the orbit over some distance. In typical synchrotrons, the corrector magnets are located every half a FODO cell.

The accelerator also needs cooling connections for components, leads for connecting magnets to power supply, wires for sensors that measure temperatures, flow, magnetic field, etc. Each of these synchrotrons would typically have a number of modular FODO cells which contain the following systems and components:

- Dipole Magnets which bends the beam path into the design orbit
- Quadrupole magnets to focus in one direction (say horizontal) and another quadrupole magnet spaced appropriately to focus in the other direction (say vertical), to provide the Alternate Gradient Focusing.
- A so-called spool piece, which contains correction magnets and houses the power leads, sensor wires and liquid helium and liquid nitrogen plumbing connections.

Figure 11.7 shows a FODO half cell. There would be a number of these standard cells in the curved sections while there would be other sections that have specialized cells specifically tailored to match the orbit characteristics of these sections. For example, as we noticed in the Supercollider ring, there would be straight sections along the way for various purposes. In transitioning from a bending section (an arc) to a straight drift section without dipole magnets, the beam would need to be prepared because the beam (betatron) oscillations in the magnetic field

Fig. 11.7 A FODO half-cell F – focusing, D – defocusing quadrupole, B – bending magnets and Sp – Spool pieces

would need to be matched to avoid dispersion (broadening in energy distribution and spreading out in space) of the particles. This is like a car on a race track. While coming from a curve to a straight section, the drivers must adjust their driving to prevent the car from being pulled apart and from careening off. This is dispersive, because different parts of the car would feel slightly different force in magnitude and direction. To overcome this, there are Dispersion Suppression (DS) cells, which use specific arrangement of magnets to keep the beamlets together.

The Interaction Region

The collision region where physics experiments take place is called the interaction region (IR). The IR prepares the beams to bring them to a sharp focus, reducing the beam diameter by several orders of magnitude to sub-millimeter size, and makes the two beams collide with each other, over a significant length of the beam bunch. The steering and focusing have to be done over a specific distance and usually there is no flexibility in the length and the path available for IRs. The arrangement of magnets to steer the beam with high precision and to focus the beam to the finest possible pencil (minimizing amplitude of betatron amplitudes while preserving beam stability) requires very advanced calculations and particle simulations. The aperture of magnets, particularly the quadrupoles has conflicting requirements – a high field quality and high radiation mean large radius, but high field strengths are more difficult to achieve with larger radii. Such focusing and guiding of the beams requires special quadrupole magnets and specialized bending magnets with extremely low error and reproducible fields, since the beam radius has to be increased first before bringing it to focus. These magnets typically test the limits of state of the art technology with large magnet aperture, strong magnetic fields (gradients) and high accuracy. Figure 11.8 shows a schematic of the interaction region. The two lines at the left and right show two beams and at the interaction point they cross. The up and down bars indicate quadrupole magnets with opposite focusing (similar to F and D magnets above), while the centered bars represent the dipole magnets. One can see that the bending magnets bring the two particle beams into the crossing situation.

The Radio Frequency System for Acceleration

The other important system for the accelerator is the RF system, which provides the accelerating voltage, as bunches of beam particles pass through. As explained before, the accelerating electric field must be high to get the maximum acceleration per pass through and yet be within limits of reliable operation (without electrical breakdowns). The frequency of the alternating electric field should be such as to be

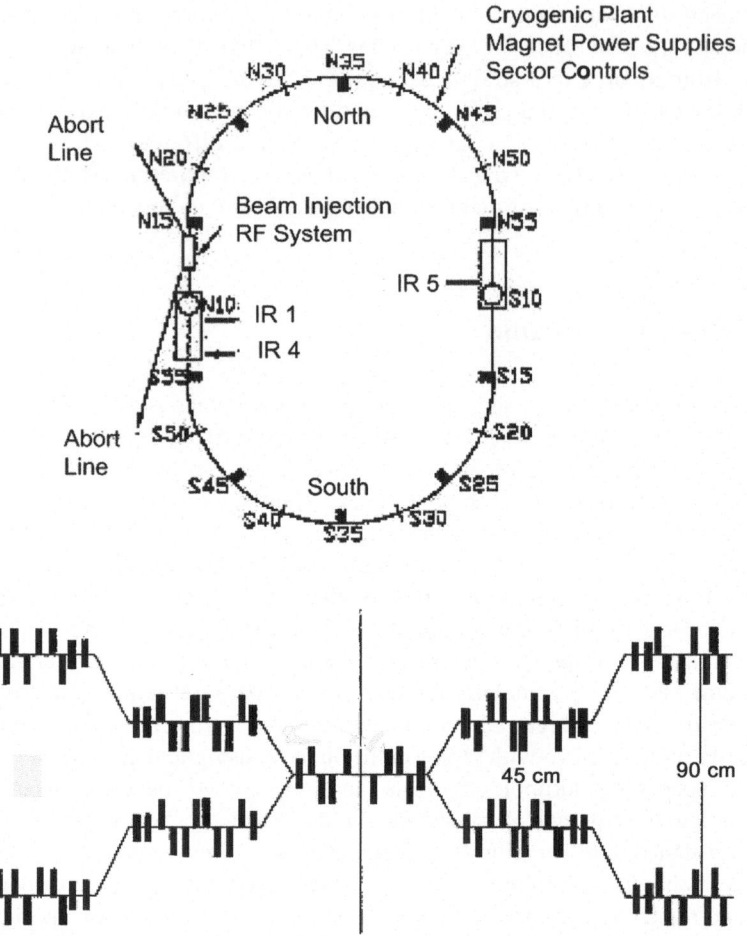

Fig. 11.8 The location of interaction regions for the SSC and the elaborate arrangement of the bending and dispersion suppressor dipoles and focusing quadrupoles (lenses) (SSCL Site Specific Conceptual Design Report, July 1990)

of one polarity when the bunch comes in and reverse as the bunch leaves. For example, for each main (collider) ring of the SSC, the 360 MHz, 2 MW RF power is generated by two Klystrons, capable of providing a peak accelerating voltage of 20 MeV. Since the beam injected from the HEB ring at 2 TeV was to be accelerated to 20 TeV in this ring, the particles would have made about 900,000 orbits during this acceleration (approximately 78 million km), taking about 300 s. During this acceleration, the magnetic fields would have been slowly raised, to keep the particles in the desired orbit.

Detectors

Last, but not least, the experimental facility has spectacularly large, cathedral size halls, housing massive detectors. These detectors have large powerful magnets to bend charged particles emanating from the collision region. As the particles traverse this region, they travel through layers of material and sensors that slow them down and detect them. The radius of curvature of the particles and how they slow down indicates the mass and energy of the particles. The size of these detectors, and the amount of wiring that goes in and out of them, would awe even the most seasoned physicist or engineer. The physics, technology and engineering of particle Detectors have come a long way from the days of phosphors, cloud chambers and photographic plates. Often, the detectors are the most visible part of a particle physics experiment, obscuring the complexity and size of the distributed accelerator system. This is also where the interesting physics experiment happens. So once the accelerator starts working reliably, it just becomes a source of high-energy particles, a commodity and all eyes are focused on the Detectors. The Detectors are described in the next chapter.

Other Physics and Engineering Systems

The above description gives only half the story. An accelerator–collider complex requires substantial diagnostics rivaling space systems. The beam has to be sampled for its nominal behavior (number of particles, bunch characteristics, cross section, etc.) and its energy and the spread in energy. The whole system of magnets, accelerator cavities, vacuum, etc. has to be monitored for these large assemblies. The beam is often brought out (using, what are called, dog legs) for other non-collision experiments such as fixed target experiments or to make a factory for producing copious amounts of exotic particles such as B-mesons. Also, when the experiment is done or the machine is to be aborted, there have to be particle dumps that absorb the other worldly energy of beam particles. Such facilities require additional physics and engineering design, technology, and fabrication.

The data acquisition, retrieval, and analysis system would be the most advanced required for any human endeavor. For example, with 40 million collisions a second, a modern day physics experiment like the Large Hadron Collider alone would generate 1 PB (petabyte or 10^{15} bytes or one million gigabytes) per year. This much data would have to be crunched down to few tens of terabytes for analysis purposes. Even this analysis, before it appears before human eyes, would have to be processed in a short time to give a good understanding of the experimental result. There would be thousands of researchers participating at the same time and the data has to be shared at high speeds. Over and above this, the system data would continuously be pouring some at a slow rate, but many at fast rates.

Fig. 11.9 The cavern for an
SSC Detector (Credit:
Freephotos)

Superconducting magnet based synchrotrons and detectors required large liquid
helium and liquid nitrogen supplies. For example, the US Supercollider's accelerator rings would have had a liquid helium inventory of a few million liters and a
similar storage capacity. An experimental collider complex requires tremendous
amount of support facilities for utilities. For example, the US Supercollider project
required 185 MW of power and an industrial cooling water requirement of 45,000
gallons per minute and similar low conductivity water and pond water
requirements. Clearly, such a large and complex system would require advanced
control and display and safety systems.

From the above, it can be seen that present day accelerator–colliders would also
occupy a large space. As a result, builders have to acquire a large piece of land, and
place the actual accelerator string in an underground tunnel. As technology for road
construction has advanced, the tunnel drilling technology also has advanced. Such
long and wide tunnels are now constructed quite easily and quickly. People working
in these accelerators take an elevator down and then use a trolley or bicycle to
navigate around the tunnel complex. The acquisition of such large area of land and
equipping the complex with electrical, communication, data acquisition and analysis, and cooling systems is a major task that accompanies the construction of the
accelerator. Figure 11.9 shows the tunnel shaft, an experimental hall cavern and the
tunnel for the Supercollider.

The particle collider experiment is the culmination of all the physics and
technologies that have come before. While there are a few additional concepts for
particle acceleration and physics experiments, the collider is the gold standard for
High-Energy Physics (HEP) experiments. This chapter has mostly dealt with
synchrotron colliders with circular orbits, but linear colliders with no bending
magnets but many RF accelerator cavities are equally feasible and in some cases,
more so. Though vast in scope of work, RF accelerated, magnet steered colliders are
here to stay, at least for several decades in the future.

Chapter 12
Particle Detector Experiments

When an accelerator delivers its promised energy and number of particles at the collision point and the particles collide with the intended intensity, the scientists would just take the accelerator and the interaction region for granted, almost like it is just a tap, only to be closed or opened. The detectors ARE the experiment, the raison d'etre for the accelerators. All eyes are on the signals from the detectors, all watching the computer screen breathlessly for that shower of "manna" from the collision point. The detectors are the citadels of physics with towering magnets that bend the phenomenally energetic collision particle to their will. The detectors are the vehicles with which physicists explore using millions of sensor components watching the particles as they wend their way through the detector assembly. The detectors are the physicists' eyes, and the control, data acquisition, and display systems are their brain assimilating the information.

So, it is no wonder that whatever the organization for the construction of the synchrotrons may be, modern detectors are invariably world-wide collaborations. Each detector collaboration has a personality of its own, with its own goals in physics discovery and its own style of achieving these goals. The knowledge that is employed in the detector is derived from the history of particle detection from the days of Thompson and Millikan, and every trick known to men is used to entice the particles to reveal their identities and nature. The detector is one of the most complex and the largest apparatuses known to man, consisting of many different types of particle detection instruments assembled into one. Below are descriptions of some detectors that particles will run into.

Particles Blowing Bubbles, the Bubble Chamber

While the cloud chamber was vapor condensing along particle tracks, a comparable but more sensitive detection method was discovered by Donald Glaser in 1952. The method is essentially based on the principle that when a liquid is superheated, that is, heated to a high temperature at a high pressure so that the liquid does not boil, and then quickly the pressure is reduced, the liquid would be ready to boil and start creating bubbles. At this point, if a particle enters the chamber, then it would ionize the atoms and the charges would seed the start of the bubbles. The bubble size and number then would grow along the tracks and become visible, fit to be photographed. So, while the cloud chamber has a supercooled vapor ready to condense into liquid drops, the bubble chamber has a superheated liquid ready to create vapor bubbles. In order to form the bubble, the liquid needs the latent heat of evaporation, and this energy comes from the heating of the liquid from the energy lost by of the particles colliding with the liquid atoms (Fig. 12.1). As in the cloud chamber, the particle tracks are bent by placing the chamber in a magnetic field and the tracks are photographed from different angles to get a three-dimensional view of the track. While Glaser used organic liquids for his chambers in University of Michigan, later experiments, such as in Luis Alvarez's laboratory, used liquid hydrogen which gives exquisite sensitivity to particles and a three-way pressure valve to allow change in pressure in milliseconds to give very accurate and short times of detection.

The bubble chamber is superior to the cloud chamber because it is more sensitive to particles produced in the high-energy accelerators such as strange particles. The much higher density of liquid compared to that of vapor makes it so. This type of detector also permitted faster photographing with a higher resolution and resetting of chamber at better than 10 times a minute, which is essential for the adequate use of the accelerators. While nuclear emulsions could record such events, the emulsions record tracks all the time and do not give an automatically triggered time window so that one gets too many tracks, making analysis difficult. The bubble chambers were an instant success and dominated particle detection for 30 years since its invention in 1953. Millions of stereo pictures were taken, half of them in CERN, and over hundred bubble chambers were built. Bubble chambers such as the

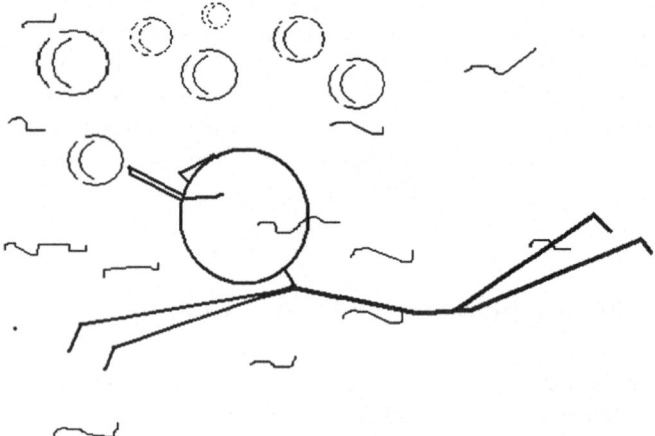

Fig. 12.1 Wow! This particle is having fun in the superheated hydrogen liquid!

80-in. bubble chamber in BNL and the 2-m diameter, 4-m long, 3.5-T Gargamelle helped mature the physics of strange particles and the family of particles on which the standard model would be based. For example, the SU(3), "the eightfold way" proposed by Gell Mann and N'ermann predicted the order of arrangement (symmetry) of baryons (neutrons, protons) and certain mesons as having the same spin and parity, but differing in a step-wise change in mass, charge, "baryon number," and "strangeness" (see Chap. 10). This arrangement predicted the existence of an omega particle. This was indeed found in Brookhaven in the 80-in. bubble chamber (Fig. 12.2). In 1973, in a most important and prestigious discovery in CERN, the giant Gargamelle bubble chamber detected the weak neutral currents (flow of particles) associated with Z bosons, a particle that carries the force of weak interactions. This discovery was the first evidence for the correctness of the Nobel Prize-winning electro-weak theory proposed by Abdus Salaam, Sheldon Glashow, and Steven Weinberg.

Glaser won the Nobel Prize for his invention. Though an apocryphal story is told that he discovered it when he was drinking beer with buddies, he did use beer once in his test, only to fill the room with a bad smell, which was not worth withstanding because the beer bubbles were not sensitive to particle tracks. The bubble chamber is no longer much in use, because of the need to photograph the tracks, which is impractical in large detectors which acquire data through electronic means and also because the process of detection is too slow.

Fig. 12.2 The Gargamelle bubble chamber at CERN (Courtesy: CERN European Organization for Nuclear Research, CERN photo Archives,Image Reference - CERN-EX-7009025)

Particle in the Cross Wire: Multi-Wire Proportional Counters

Far back in 1908, Hans Geiger invented a very effective way to detect radiation (Chap. 3). Almost all particles – alpha, beta, gamma, and other particles – ionize the medium they travel in. The Geiger Counter remains nearly unchanged in concept, in which a metal tube connected to the negative of a DC supply has a central wire at positive potential. Typically, a mica or a similar window seals the ends of the tube. The tube has an easily ionizing gas such as argon. When a particle arrives in the gas, it ionizes atoms, releasing free electrons. These electrons are attracted to the positive wire and cause a current to flow, which is detected as a voltage drop across a resistor. A quenching gas in the tube mixed with argon stops the process quickly so that one gets a short pulse of current and "a count of radiation." Walther Mueller refined this device to improve the sensitivity and accuracy of detection. The put-putting and crackling counter, one often sees in real and movie situations, is really this device.

A variation of this is the Proportional Counter (Fig. 12.3), in which the applied voltage is high enough and pressure is low enough that the free electrons created by the incident particles themselves gain energy and collide with atoms to create an avalanche of ionization, releasing more electrons. The electric field is shaped (see Fig. 3.2) such that it is relatively weak away from the central wire. This way, the

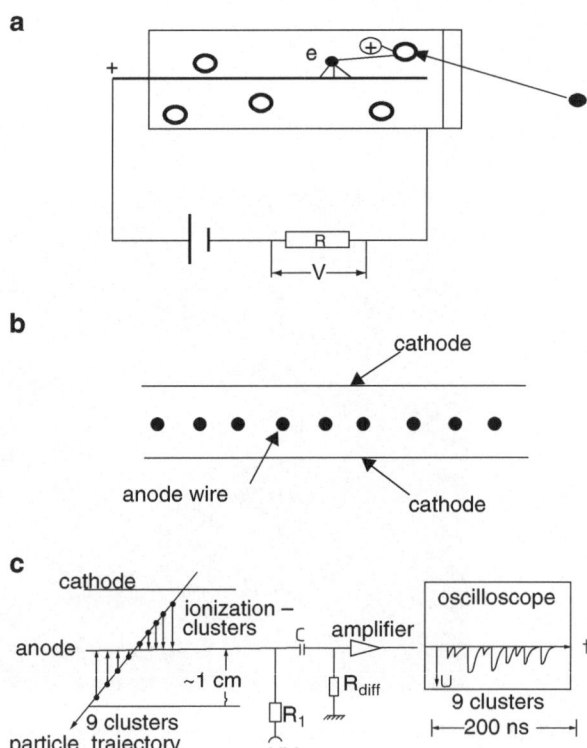

Fig. 12.3 Principle of proportional counters, the *top left* figure (**a**) shows the production of an avalanche, the *top right* figure (**b**) shows the arrangement for a multi-wire proportional counter. The *bottom* figure (**c**) shows the electronic detection of these events so that time coding can be obtained

avalanche multiplication occurs only close to the wire and is insensitive to where the primary electron is born. There is a delay in time depending upon where the electron is born (see also Chap. 3). A particle arriving produces several electrons as it slows down, and the larger the energy of the particle, the farther it travels, and larger is the number of primary electrons and therefore secondary electrons. The amount of charge flowing through the electrode is proportional to the initial energy of the particle, hence the name – "proportional counter." The time the pulse arrives is related to the delay and, therefore, the location the electron came from. As the particle proceeds along the medium, the number of ionization events is recorded as a function of time. The figure shows an example of signal peaks corresponding to 9 avalanches (clusters). The high-tech, modern version of the proportional counter revolutionized detection of particles by bringing it into the electronic age. While the proportional counter tells about the arrival of a particle with a specific energy, it did not provide a track. One had to depend on a bubble chamber to see the tracks. However, in 1968, Georges Charpak invented the multi-wire proportional counter (drift chamber). In this, an array of wires is placed so that a particle location is determined as a function of time by the signals from each of the wires (see bottom Fig. 12.3). In more elaborate counters, another array, adjacent to the first array, pinpoints the position of the particle even more accurately, from the time of arrival of the signal at the two different wires. The Multi-Wire Proportional Chamber (MWPC) is placed in a magnetic field so that the charge-to-mass ratio can also be determined from the curvature of the track. The big advantage of the MWPC and its sister devices is that now the signals are electrical and the position is coded by the grid laid down by the wires. This way the particle path can be imaged and tracked on the screen just like a culprit being tracked in traffic grid by the police. Data acquisition systems and computers handle all the data processing and one can even select the type of tracks one is looking for. The wires are placed millimeters apart with a precision of tens of microns and with this, the particle position at any instant can be tracked to within 50–300 μm.

Highly sophisticated versions of the proportional counter-wire chambers are now used. In the Drift Chamber version, the field is shaped to be very uniform away from the wire avalanche region, and as a result, the arrival time of the ionization pulse tells the distance from the anode wire more precisely, and the relative signal between two ends of the same wire gives the position along the wire. Two wire arrays oriented perpendicular or at an angle of 45° can resolve the overlapping signals of two simultaneous ionization events due to two particles. More complex multiple wire arrays have grids on which the drifting electrons are first collected. Then the grid is gated open to pour out the electrons that create avalanches, and both electrons and ions are detected.

A variation of the same principle is used in which the avalanche produces a plasma, and a spark occurs between wires maintained at a high voltage. This device is called a Spark Chamber. First, a scintillation counter detects a particle and turns on the high voltage on the wires and then, as the particle traverses between electrodes, it creates a plasma spark. Once the spark is detected, the high voltage is turned off.

Scintillation Detectors

When particles or gamma rays pass through certain materials, the atoms of the material absorb particle energy through excitation and ionization processes. When the atoms return to ground state or the ions recombine back to atoms, light is emitted. This flash of light is detected to "count" the particle, and the detector is called the scintillation counter. While the scintillation light emitted is often in the visible range, some materials scintillate in the UV range. In the latter case, a fluor layer or doping is added which absorbs the UV light and reemits (fluoresces) the radiation at a more detectable wavelength such as blue. A good scintillation counter has a fast response to particle event, quenches fast, and does not reabsorb the emitted light. Plastics such as lucite, crystals such as anthracene, and inorganic materials such as sodium iodide (doped with thalium) make good detectors. In scintillators such as the plastics, a fluor layer also helps in preventing reabsorption by shifting the wavelength of light. In all crystalline and noncrystalline detectors, dopings and additives tune a detector to a specific need.

In modern detectors, light is detected by photomultiplier tubes (PMT) or photodiode arrays. Light falls on the PMTs or is coupled into the PMT window through optical fibers. The photomultipliers (Fig. 12.4) have a surface which emits electrons when a light quantum hits it, because of the photoelectric effect. An array of electrodes (dynodes), while multiplying the electrons with secondary electrons, also accelerates the electron stream and a final anode detects a pulse corresponding to a count of light emission. Photodiodes are also sometimes used because of their compact size and efficiency, but these tend to be noisier. While the performance in time resolution and dead time are comparable to gas-filled detectors such as proportional counters, and the signals are also proportional to the particle energy, the efficiency of detection by the PMT is poorer because of lower quantum efficiency (electron emission event per photon). Energy resolution is also three to four times poorer than that of gas-filled detectors. These also tend to age with radiation damage. But these are convenient devices and are commonly used.

Solid-State Detectors: Silicon Detectors

The good old Silicon, the omnipotent and omnipresent element, is also one of the most useful detectors. The silicon detectors are the so-called p–n diodes, a common type of device in electronics application. When radiation (particle) passes through the diode, an electron may get knocked out of the atoms or excited into the "conduction band," where electrons freely roam about as if a gas. This creates a "hole" in the lattice, which is similar to an ion. The movement of these electron-hole pairs near the junction between the p and n materials causes a current to flow, which is detected as a pulse. Aided by the demands of the consumer and computer industry, the silicon detector can be manufactured with nano precision, which

Fig. 12.4 A photomultiplier
tube

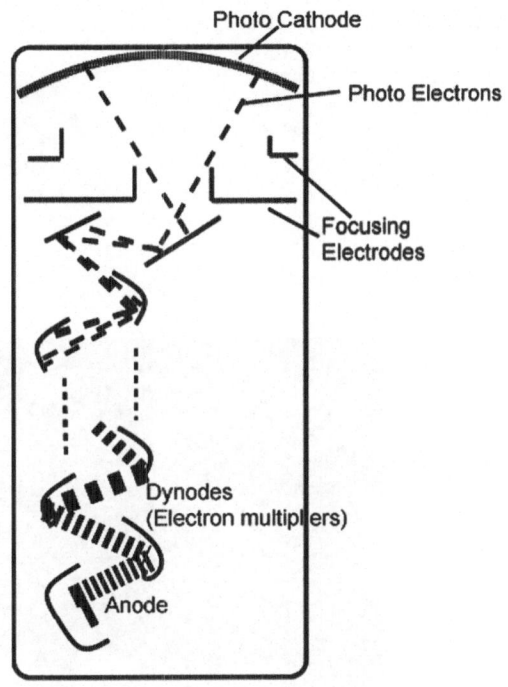

translates into accurate determination of the particle position. Because the energy of excitation is small, the energy resolution is high and silicon detectors give response times limited only by the speed of the read-out electronics. The sensitivity is also limited only by the noise in the read-out electronics. The only drawback of the silicon detectors is that they require to be cooled to minimize thermally induced leakage current and aging due to radiation.

The modern silicon detectors have become so small that they are now deemed pixels, like pixels on a computer screen. Because of their small size and their fast time response, these are the only detectors suited for recording particle tracks close to the particle collision points in a collider. Here, a large number of closely packed particles emerge and many of them would decay away before they get into the other types of detectors. With their fine granularity and thickness, these "Tracker" detectors form the front line and are placed right next to the collision point to locate and record the presence of such particles with speed and precision as they cross each pixel. In a modern detector, there are millions of such detectors in the form of arrays, and each pixel has its microchip read-out electronics. The composite of the pixel and read-out electronics forms an integrated package to give a fast time response. The signal, like many other signals, is brought out by optical fibers. To maintain the compactness of the assembly, the electrical power connections are usually thin ribbons and the whole assembly is cooled with coolants such as fluorocarbons. The structures holding these dense assemblies have to be designed

Fig. 12.5 A silicon pixel
detector arrangement, where
a layer of sensors is directly
soldered on to readout chips
with solder bumps. (Courtesy:
University of Tennessee
Knoxville, USA)

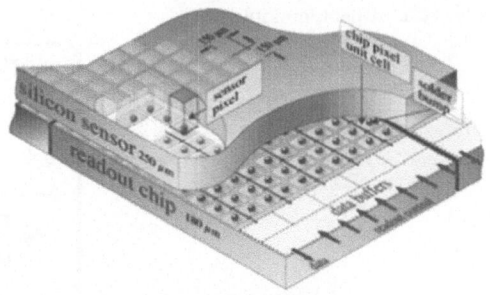

Fig. 12.6 3 Disks of the ATLAS pixel detector with the detector electronics (on the end cap) – the complete assembly has 80 million pixel detectors in an assembly about only 1.5 m in length and 0.5 m in diameter. (Courtesy: Lawrence Berkeley National Laboratory, Physics Lab View, May, 19th, 2006.)

so that they do not block particles, and often carbon fiber material is used for this (Figs. 12.5 and 12.6).

Cerenkov Counters

In 1934, S.I. Vavilov and P.A. Cerenkov reported that gamma rays moving through a solution gave rise to two radiations, one that was well understood to be just luminescence or fluorescence where the medium just absorbed the light and remitted light in the visible region. The second type of light, the Vavilov–Cerenkov radiation, arose from the fact that these gamma rays created the so-called Compton electrons that moved at a speed faster than the speed of light in the medium – are superluminal (The electrons still do not violate the theory of relativity, because the speed is less than the speed of light in vacuum.). The radiation is similar to a shock wave and is produced because the particle polarizes the medium. When a charged particle moves through an insulating medium, the moving particle's electromagnetic field affects the medium and the medium is polarized with displacement of electrons in atoms. After the particle has passed from the vicinity of the atom, the electrons in the atom return to their equilibrium position and in that process emit a light. In normal circumstances, at low speeds of particles, the light is emitted incoherently and the light waves interfere destructively with no net light output. But when this disturbance occurs at a speed faster than the light being emitted, there is no full cancellation and a net light output is observed. This radiation has a threshold of $\beta n = 1$, where β is the ratio of particle velocity v to that of light c, and n is the index of refraction. The phase velocity of light is v_n (speed of light in the medium) $= c/n$, the light is emitted at an angle given by $\cos(\theta) = v_n/v = 1/(\beta n)$ (this is similar to the sonic boom which also has this dependence). The light is emitted over all wavelengths, but the intensity is proportional to the frequency of light (or inverse of the wavelength) and is cut off at about the UV wavelength. In a light water nuclear reactor, the blue Cerenkov radiation can be seen surrounding the fuel rods (Fig. 12.7).

Hence, this mechanism is a detector of the particle velocity. A Cerenkov detector may be (1) a threshold counter – the mere presence of this radiation with a characteristic frequency dependence shows if the particle exceeds the speed of light in the medium, so every such particle can be counted; (2) a ring imaging Cerenkov detector (RICH) measures the angle of the radiation (the cone of radiation which casts a ring of glow on the detector) to determine the speed of the particle. If the momentum of the particle is known from other measurements such as a track in a magnetic field, then the mass can be inferred from the velocity measurement using the Cerenkov detector. By selecting the medium, one can get the best intensities and preferred radiation cone angles. The Cerenkov radiation detectors are now part of any large detector experiment, more often than not, the MWPCs are embedded in the Cerenkov detector medium.

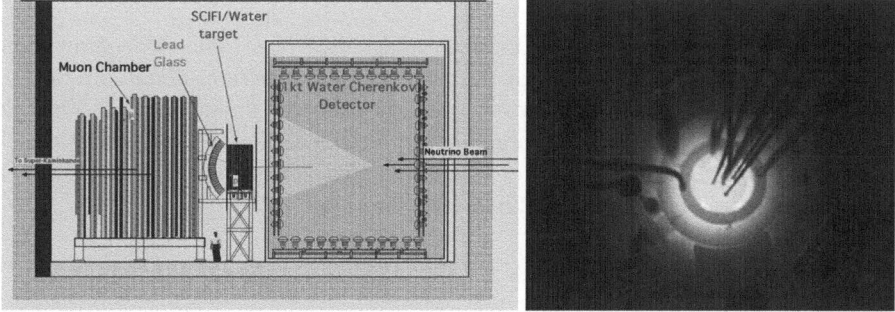

Fig. 12.7 *Left*: The Near Detector-Neutrino oscillation detection experiment in KEK, Japan, with a Ring Imaging Cerenkov counter, (Courtesy, KEK/JAEA, Japan) *Right*: Cerenkov radiation from a nuclear fission reactor core (Courtesy: Oregon State University, TRIGA Reactor, Corvallis, OR, USA)

The DELPHI detector in the Large Electron Positron (LEP) collider used a RICH detector for the first time. Chambers containing liquid and gas fluorocarbons working as Cerenkov medium were used with a specific goal of distinguishing between K and π mesons with momentum corresponding to about 30 GeV. The decay of Z0 into + and − muons was clearly seen from the angle of emission.

Calorimeters

A particle can be measured by stopping it on its tracks in a medium and converting all the kinetic energy (finally) to heat or some form of measurable energy, just like in the college physics Joule experiment. Alternatively, the particle of interest can be made to interact with a medium and create secondary particles (showers of other particles), which are stopped in a different medium and the energy of the secondaries is measured using detectors. The first one, homogeneous calorimeter, gives the bulk (integrated measurement). The second one, the sampling calorimeter, can be constructed with thin layers of alternating showering and calorimetric materials so that by the response of the individual segments, the tracks can be determined and the rate of energy deposition can be measured with detectors as a function of track distance. (One is reminded of Carl Anderson's experiment with a lead plate to identify the positron). However, the drawback in the sampling calorimeter is that the first medium that is supposed to only create the secondary showers also absorbs some amount of energy and that goes undetected. This loss has to be separately estimated *a posteriori*. The energy and signal resolution of these detectors are only limited by photon statistics and fluctuations in particle path or leakage transverse to the beam. Typically, one can achieve an energy resolution of less than 10% for MeV-type energies. If a calorimeter is constructed well and is "hermetic," it collects nearly all the secondaries and measures all the energy. The calorimeter is but a combination of the stopping/showering medium and scintillation and Cerenkov detectors. By immersing the calorimeter in a strong magnetic field, the effective path length is increased and particle curvature in the magnetic field is recorded.

Calorimeters are also classified as electromagnetic or hadronic, depending upon whether they absorb energy through electromagnetic interactions (Coulomb forces, atomic excitations, ionizations, etc.) or through hadronic interactions, in which a hadron, say a proton, passes close enough to a nucleus that it applies a "strong" force to knock out a neutron or a proton (hadrons) from the nucleus and then there is a succession of these events called hadronic showers.

EM Calorimeters

Electromagnetic calorimeters are not sensitive to hadrons and exclude hadronic information, giving information on photons, electrons, and positrons in the range of MeV to hundreds of GeV. Electromagnetic calorimeters use crystals such as lead glass or sodium or cesium iodide. The non-hadronic particles collide and create secondaries (by pair production, ionization, and bremsstrahlung and Cerenkov radiation). The secondaries also create tertiaries and higher generations, until finally only low-energy photons or electrons are left which are absorbed (energies are absorbed) in the materials. The materials then reemit the energy as light, which is what is detected, using photodetectors. The sampling calorimeters use alternating layers of an inactive material such as lead and active materials such as scintillators. Liquid argon is also used as the calorimetric medium, because it acts as a scintillating medium.

Hadronic Calorimeters

Since hadronic showers only occur in the less probable "close encounters" with the nucleus, they require thick and dense materials such as iron and lead, to create signals. While all materials produce electromagnetic or hadronic showers, one wants a material that requires minimum interaction length. Table 12.1 shows the mass of a material required per unit area of the shower. One can see that lead is best for electromagnetic calorimeter. While hydrogen appears to be superior to metals for hadronic calorimeter, being a gas it would need a large volume. Iron or lead would again be the choice to reduce the size of the calorimeter. All the same, one can see that hadronic calorimeters would require 10 times more thickness than EM calorimeters. In order to provide full coverage, the hadronic calorimeters fully surround the interaction region with a "barrel" and two-end caps. The hadronic calorimeters can also be a sampling calorimeter as in the CERN CMS detector, which is a combination of inactive 5-cm-thick brass and 4-mm scintillator sections (see Fig. 12.8). In the ATLAS detector, the calorimeter consists of iron and scintillator crystals, and another inner calorimeter with liquid argon.

Hadronic calorimeters are specially suitable for characterizing "jets," where a cone of secondaries sprays out of the particle path. The source and nature of these jets are very complex, but all the same provide an important additional diagnostic of the particle collision. In the CMS detector in Large Hadron Collider, the central end plug is made of iron and quartz in order to be able to withstand the high radiation dosages from the jets. While hadronic calorimeters also have ionization and scintillation processes, the strong force interaction makes for additional

	Material of the calorimeter	EM (g/cm^2)	Hadronic (g/cm^2)
Table 12.1 Comparison of different choices for electromagnetic versus hadronic calorimeters	H_2	63	52.4
	Al	24	106
	Fe	13.8	132
	Pb	6.3	193

Larger values are better, but the volume required would limit the use of hydrogen, and handling and dust (such as toxicity of lead) also need to be taken into account

Fig. 12.8 CMS hadronic calorimeter detector end cap of brass and NaI scintillators (Credit: Compact Muon Solenoid experiment, CERN, European Organization for Nuclear Research, Image Reference CMS-PHO-HCAL-2002-011)

processes such as excitation of a nucleus, emission of a proton or neutron, decay into neutrinos and muons, and recoiling of nuclei. These strong interactions which can cause 30–40% energy transfer from the particle being observed do not result in light signal and, therefore, are not detected by the photodetectors. Therefore, this is estimated or measured separately.

When we get to the outer layers of these calorimeters, the only particles that travel that far are muons, because they interact least with materials. So when one detects particles (light emitted through shower mechanism) at this distance, one is sure that these are muons. Therefore, the end calorimeters and the calorimeters at the outside of cylindrical detectors are called muon calorimeters. In the Fermi D0 and the CERN ZEUS detector, the muon chambers are made with uranium.

The final detector assembly is an enormous device, consisting of a central tracker, for example, made of silicon pixel, special Cerenkov chambers with MWPCs embedded in them and large calorimeters. All the detectors are immersed in a high magnetic field created by enormous magnets, which have finger- to fist-sized superconductors, some of them specially built with specific science and technology. The size of the detector (the largest being 30 m) might rival a building and certainly would be the size of a nuclear reactor. These detectors are located in cavernous halls and if one chanced upon seeing it under construction, one would see technicians, professors, and researchers perched on high cranes diligently wiring or checking massive ropes of fiber optic cables or a complex cascade of cooling pipes. In a large experiment like the LHC, the volume of the detector would exceed the volume of the accelerator ring. While the accelerator ring's complexity is that of the

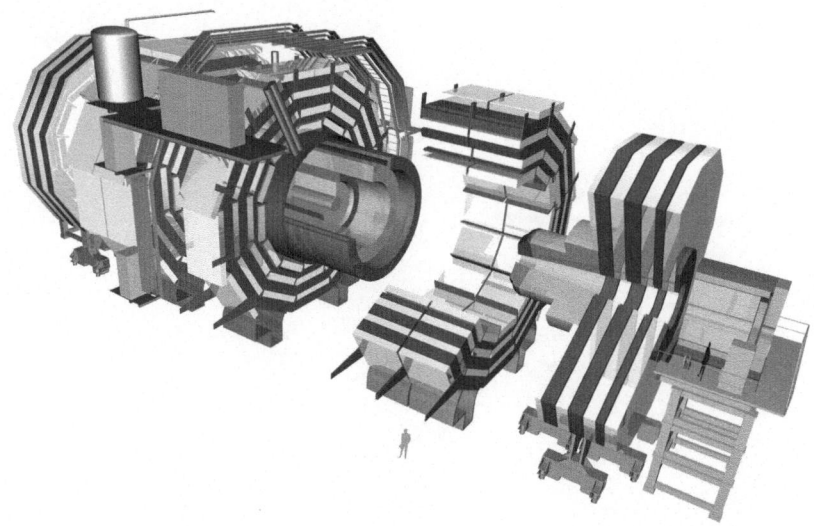

Fig. 12.9 Sections of the CMS detector in LHC (note the size of a human being) (Credit: Compact Muon Solenoid experiment, CERN, European Organization for Nuclear Research)

repeated state-of-the-art components, the detector's complexity is the simultaneous housing of millions of fragile parts which are also state of the art. The magnetic volume and field are very high. In fact, the superconducting supercollider even offered to use the GEM detector for tests on superconducting energy storage. Figure 12.9 shows the schematic of the CMS detector in the Large Hadron Collider project. One can see the relative scale of the device in comparison with a person.

Detectors are the pride of physicists, with the detector collaboration akin to a society, with its roving postdoctoral researchers, University professors, laboratory physicists, engineers, technicians, consultants, architectural advisors, spokespersons, industrial providers, and fabricators and the army of analysts. Even within the same facility, there are often several bruising competitions between two detectors, because after all, the prize is the understanding of nature itself and the trophy may be the Nobel Prize in physics.

Chapter 13
The Snake Charmer: The Large Hadron Collider

In the late 1970s, physics had come out of many confusing conundrums and had entered into a regime of physics, in which there was a greater precision in the questions that were asked. It was becoming clear that cosmological questions relating to the ultra-large sizes of space, massive objects, and eons in timescale could only be understood by investigating the world of ultra-small, mass of fundamental particles and finest of time scales. In order to understand the creation of galaxies, stars, and planets; the process of generation of the attribute of mass which gives rise to gravitational forces; balance between particle and antiparticle; the presence of dark energy that is causing the Universe to expand; how radioactive processes allowed galaxies and stars to evolve; etc., need an understanding of matter, forces, and energy at particle level. On the flip side, an understanding of the conditions of the Cosmos that existed prior to creation of particles would give clues as to the mechanism of such formation. The unity of the large and the small is similar to the ancient understanding of unity and self-sufficiency, expressed in the symbol of a snake, the Ouroboros, with its tail in its mouth. Plato was prophetic when he described such a creature as the original inhabitant of the earth, because it was self-sufficient (Fig. 13.1).

In order to examine this profound connection between the large – the Cosmos, and the small – the fundamental particle, and thereby develop a complete story, the particle accelerators would need to be in the multi-TeV range. The viability of the Standard Model (SM) and an understanding of what lies beyond can now be garnered only with very large experiments peering into the very small. So, in 1978 and 1979, the International Committee on Future Accelerators met and discussed a 20-TeV proton beam on 20-TeV proton beam collider. The 1980s spawned the idea of two colliders.

R. Jayakumar, *Particle Accelerators, Colliders, and the Story of High Energy Physics*, 199
DOI 10.1007/978-3-642-22064-7_13, © Springer-Verlag Berlin Heidelberg 2012

Fig. 13.1 The Ouroboros
(snake eating its tail)
enclosing life and death cycle,
drawing by Theodorus
Pelecanos, Synosius,
alchemical text (1478)

The Superconducting Supercollider Laboratory

In the Snowmass (USA) workshop, held in 1982, a large leap forward in High Energy Physics was proposed. Following that, the US High energy Physics Panel initiated the "Supercollider" project in 1983 and a National reference Design Study too was begun, which culminated in the proposal for a 20 TeV–20 TeV proton collider with a luminosity of 10^{33} cm^2/s. After extensive reviews and a Presidential Decision by George Prescott Bush, in 1988, the Superconducting Supercollider laboratory was created in Texas, under the directorship of Roy Schwitters. Until that time, the project conceptual work had been led by Maury Tigner, with optimistic expectation of the beam performance particularly with a 40-mm beam tube aperture. In 1990, the site-specific design became more complete and also more conservative when 107 turns around the ring were simulated using data from HERA, the first superconducting synchrotron. This simulation resulted in an increase in the dipole aperture from 40 mm to 50 mm in order to improve field quality. There was an associated cost increase in development and construction. The footprint of a 54-mile-long ring was established and 16,000 acres of land were acquired. By 1992, a large number of state-of-the-art superconducting magnets, both dipoles (15 m long, 6.7 T) and quadrupoles, were built and demonstrated at an unprecedented pace of innovation, engineering, and material development at Fermilab, BNL and LBL. In Superconducting Supercollider Laboratory (SSCL) itself, superconducting magnet design, fabrication, and testing capabilities were fully developed in a very short time. In an impressive demonstration of the capabilities of a fresh new laboratory with an international team of driven engineers and physicists (many of whom were new to the accelerator field), superconducting magnets, corrector magnets, spool pieces, and instrumentation were designed and fabricated. In a display of how quickly a motivated team could achieve, the Interaction Region quadrupoles with large apertures and an unprecedented 240 T/m strength were designed and fabricated practically flawlessly and tested successfully. Cryogenic and structural facilities including digging of 15 miles of tunnels were underway by 1992. In parallel, aided by new capabilities in particle simulations, innovative designs of the Interaction Region were developed, which would help future colliders. Two massive detector programs, the SDC and GEM, very similar to the present LHC detectors, also got underway. Overall, the sweat and intellect of 2,000 physicists and engineers, over 300 of whom were from foreign countries, were successfully creating the largest ever enterprise in physics. It was clear that there were no serious physics or technical obstacles to achieving the 20-TeV energy and the 10^{33} cm^2/s luminosity, except for some adjustments in parameters such as an increase in FODO quadrupole aperture. The cost and schedule of the project with contingencies were approved by two government-led panels. But by this time, the cost estimate for the project had doubled to over 9 billion dollars, some of it related to simply inflation due to delays in the approval of the project and the project-funding profile. Part of the cost increase was also toward preservation of the environment and the green field concept promoted for the Fermilab under Robert Wilson in the 1950s. There was also some nervousness about the aperture of quadrupoles and hardware to absorb the synchrotron radiation from orbiting protons. There would be additional cost increases for this.

The state of Texas itself contributed \$1 billion and the profile of federal funding was established with a goal of full operation around the turn of the century. Additional contributions from Japan and other countries were sought to cover at least part of the cost increase. While Japan was mulling over this contribution, the mood in US congress was not conducive for projects that could be cut with no political cost. Under the ferocious deficit reduction zeal, created and fueled by the conservative US speaker Newt Gingrich, a cost increase was just the weapon the opponents of the project in Texas and opponents of big science needed. The media publicity on the project management also did not help. After an expenditure of \$2 billion, after the full and complete development of each of the major components of the machine and after fully 1/3rd of the tunnel had been built well under cost and ahead of schedule, the US congress, in a demonstration of their preference for politics over priorities, voted in 1993 to terminate the project. The change of US presidency also did much to let the project die.

The turbulent days of SSC were well captured by the New York Times (March 23, 1993) correspondent Malcolm Browne who wrote about the SSC director:

> The project's 48-year-old director, Dr. Roy F. Schwitters, is racing along an obstacle-strewn course – beset by technical difficulties, the insistent demands of fellow scientists, and the opposition of powerful critics in Congress, the public and some scientists. . . .Dr. Schwitters, whose impeccable manners and gentle demeanor contrast with his harrowing life, spends much of his time traveling. The Oldsmobile he drives around Ellis County to various sites is equipped with a radio-operated trunk to speed access to his hard hat, safety glasses and technical instruments. Every week or so he visits Washington to testify, lobby politicians and officials, and seek political support for the supercollider. Meanwhile, he makes hundreds of technical decisions each week, for which he has to read reports, listen to subordinates, inspect complex gadgets and study.

Many would justify this termination as an appropriate measure because of the cost increase and poor management. But a proper understanding of the priorities of an advanced nation would have resulted in a reorganization of the project without giving up this valuable project which promised much by way of technological, intellectual, and sociological progress. In reality, the drama of the collapse of Soviet Union and the eastern block, and the end of cold war in the 1980s had laid the foundation for this ending. It is well known that many in the US executive and congress do not vote in favor of big science to advance the science. Instead, most had voted and vote in order to support military technology and to sustain the Cold War. Opposition to a project had little to do with cost overruns, particularly when there were no technical uncertainties. Much larger cost increases, averaging 40%, had been incurred in other projects including Space programs, weapons programs, and other civil construction programs (the Big Dig in Boston is a clear example), but had not been cancelled. In fact, with an identical project which also had the billion dollar tag, Robert Wilson was not only given the funds for the Fermilab, but was asked to come back for more if that could be spent. At that time, high-energy physics was considered to be necessary for developing military superiority and to give bragging rights to the cold war warriors. The power and influence of the military–industrial complex and electoral politics in the USA favor this disparity. It also did not help that the SSC was the first one to arrive at a budget rigorously and included verifiable estimates for the cost of every personnel, every contract, every operation, and the overheads. The accounting and reviews also followed more strict procedures, and nothing could be and was done "under the radar."

Ultimately, as John Peoples, former Director of Fermilab, put it, "When something bad like this happens, everyone has a hand in it." For nearly two decades, the ghost of SSC has been walking the flooded tunnels in Waxahachie, Texas, and haunts the minds of physicists and engineers and made a lot of people cynical. One disgusted Waxahachie town official, N.B. "Buck" Jordan, had this piece of advice for any American town that might be tempted to be a center of major science: "If there was ever anything else like this that came along," Jordan says, "Get the money up front." But the Large Hadron Collider (LHC) has put a smile on faces and brought back hope.

The Crowning of the Large Hadron "Supercollider"

In 1989, as SSC was ramping up in the USA, CERN in Europe was celebrating its successes with the Super Proton Synchrotron (SPS) and the Large Electron Positron (LEP) collider. After a few months of informal consultations between scientists, a workshop in Lausanne addressed the question of the feasibility of building a hadron collider with TeV type energy. 2 days of intensive discussion were held at CERN itself. LEP infrastructure had always been considered for this type of collider and, therefore, the 27-km tunnel and the final energy would determine the superconducting magnet requirements. Based on physics discovery expectations, particularly with the goal of finding the Higgs Boson, a 9 TeV on 9 TeV machine was proposed, which required the development of 10 T magnets, a demanding task. The Lausanne workshop summary states:

> To be competitive, the LHC has to push for the highest-possible energies given its fixed tunnel circumference. Thus the competitivity lives or dies with the development of high field superconducting magnets. The long gestation period of LHC fits in with the research and development required for 10 T magnets (probably niobium-tin), which would permit 10 TeV colliding beams. The keen interest in having such magnets extends into the thermonuclear fusion field, and development of collaborations in the US, Japan and Europe feasible.

Since, at the time, the machine would be competing against the SSC which had a much larger energy, it was proposed that the machine should compensate with an increased luminosity, and the luminosity target was set at 10^{34} cm^2/s (10 times SSC). It was clear that though proton–antiproton collisions would have cleaner signals for new discoveries, the luminosity goal could not be met and the machine would also become complex requiring storage rings for antiproton beam bunches. Therefore, the collider would be a proton–proton collider (For energies greater than 3 TeV, the disadvantage compared to that of antiprotons is much less.). With these targets set and the tunnel chosen as the LEP tunnel, the rest of the design followed. The choice of the LEP tunnel forced not only the strength of the magnets (the bending radius was restricted to the LEP tunnel radius for a given energy), but also one other design choice because of the tunnel width – while the two counter-rotating proton beams would be in separate tubes enclosed by separate coils, the two magnet coils would be held in the same iron yoke – outer shell enclosure (see fig below).

Soon after, in 1992, CERN council declared its commitment to build the Large Hadron Collider (LHC) and approved a program of development of superconducting material and engineering development. Despite that, the bureaucratic machine would only move at a slow pace. After many years of wrangling and delays, on 16 December 1994, on the final administrative hours of CERN's eventful 40th anniversary year, the new Director General Chris Llewellyn-Smith got the thrilling news of the unanimous approval, by the CERN's governing council, for the construction of the LHC collider in the LEP tunnel. After the decision had hung in the balance until the last possible moment, CERN received the best 40th birthday present it could have wished for – a unique machine which will provide a world focus for basic physics research. Carlo Rubbia, the Nobel Prize-winning physicist and Director General until the previous year, had worked tirelessly for a collider of this kind. In his unashamed attempts, he was even mocked by Robert Wilson of the USA as "the jet flying clown" on what is now known as the "Tuesday Evening Massacre." With the untimely demise of the SSC, the LHC inherited the mantle of "Supercollider."

The project was delayed due to technical problems with magnets and costs mushroomed from an initial estimate of $2.5 billion to $4.6 billion, and the total cost including all aspects of operation and personnel would be $9 billion. Despite the cost increase and schedule delays, the support from the governments and the CERN itself was unwavering, and today, LHC is a functioning reality already delivering experimental results. This is in clear contrast to the short-sighted cancellation of the SSC in the USA, and Europe can be proud of its visionary decision to keep its eyes on the prize.

This massive machine was commissioned with the arrival of first particle bunch in August 2008 and the first beam was circulated in September 2008. But a magnet quench resulted in damage to about 100 magnets and the machine was shut down for several months for investigations, repair, and changes to prevent further such quenches. In November 2009, proton beams were again circulated and since then the machine has not looked back. Within days, LHC became the highest particle energy machine achieving 1.18 TeV, and on 30 March 2010, the machine reached the milestone of 3.5 TeV on 3.5 TeV, a high watermark for physics, catapulting the machine into the research phase. Already exciting physics results have started coming in (see below).

At the end of April 2011, the world was abuzz with rumor that the much-anticipated Higgs Boson had been found at the LHC ATLAS. Even the unconfirmed rumor, later proved untrue, was published widely in the press, indicating how the public is excited about new physics.

At the Threshold

The LHC, the largest physics experiment ever built, is constructed to answer questions, many of which are well posed and some which are speculative. The LHC will definitely advance or cut down several existing theories and theoretical models and, at the same time, give clues to the viability of less verifiable theories. It is certain that LHC will be exploring the horizon of knowledge and frontiers of physics, and the best of the scientists will see many results with child-like awe and semi-understanding of lay people. Some of the goals of LHC are listed below and these are but a general overview. In reality, LHC will provide many detailed answers to a wide range of questions and fill in gaps while advancing physics and technology. The LHC machine is as beautiful and esthetically pleasing as it is complex, and readers would be delighted with thousands of archived pictures at the CERN website.

Operational Experience and Benchmarks

In the first few years, simple operational experience in this biggest ever machine will provide great learning in accelerator and detector physics and technology area. With the unprecedented energy of collisions, LHC will generate very large number of events and vast amount of data. Much will have to be learnt in the process of establishing benchmarks for what is old and what might be new so that new particles and phenomena can be identified without ambiguity.

Confirmation of the Standard Model

As we saw, the SM has an open question on the origin of particle mass and variation of mass from particle to particle. It proposes a Higgs field intermediated by the Higgs boson to confer the mass. If the Higgs particle exists, LHC will find it and confirm the Standard Model. LHC will acquire the full ability to detect Higgs bosons in a few years after commissioning and a year after achieving full luminosity, when the potential Higgs production rate is expected to be one every couple of hours. It is expected that the theoretical, modeling, and data analysis developments will be instrumental in ascertaining the discovery and the properties of the Higgs boson. However, this confirmation of the SM, though a key step, will not tie up several remaining loose ends. LHC will be able to investigate all these areas and will probably resolve many of these.

- If the Higgs boson confers mass, then how does it get its own mass? Its interaction with itself in conferring its own mass is not something that can be addressed by the SM. Only extremely fine-tuned and fortuitous quantum cancellations can avoid the singularities and infinities. Though a similar problem was solved for the electron interacting with itself in Quantum Electrodynamics (QED), the SM does not have the tool to address it.
- Also in the SM, the three forces, electromagnetic, the weak force, and the strong force, come closer at very high energies, but differences remain and the three never truly unify to give a Grand Unified Theory (GUT). The Standard Model cannot address unification of gravitational force with the particle forces.
- The Standard Model cannot account for the dominance of the observed range of matter-to-antimatter ratio (See below).
- It is now known that only 4% of the matter (mass) in the Universe is recognized by astronomical observations. Astronomical observations show that galaxies spin much too fast, indicating a large "missing" (meaning unaccounted and unseen) mass of dark matter, and star motion and refraction of light indicate that the stars and galaxies travel and move around a mass of dark matter invisible to us. This had been noted as early as 1934 by Fritz Wicky. It has been well known that the Universe has been expanding since the Big Bang birth of the Universe. But recent astrophysical observations show that the expansion is accelerating. This implies that the Universe is also uniformly permeated by a form of dark energy which causes a repulsive force, causing the acceleration. It is speculated that 23% of the matter is dark, not emitting any form of radiation or particles, and 73% of the matter is actually dark energy. The SM has no handle on this. The speculation that some of this mass may be that of the neutrinos cannot be confirmed by the SM and associated experiments, since neutrino mass is within the uncertainties and errors of energy balance calculation or measurement. Most of the dark matter is known to be non-baryonic (that is, not made of protons and neutrons).

Confirmation of SUSY

In 1966, a theory of supersymmetry was proposed by Hiranori Miyazawa and later refined by J. L. Gervais and B. Sakita, Yu. A. Golfand and E.P. Likhtman, D.V. Volkov and V.P. Akulov, and J. Wess and B. Zumino. The supersymmetry model SUSY proposes companion particles to the SM particle set – the superparticle set or the sparticle set (e.g., squark). Each sparticle is different from the corresponding particle by half spin so that boson particles would have a fermion companion and vice versa. With a further symmetry breaking from SM, the SUSY predicts much heavier (100 GeV to 1 Tev) companion sparticles than the companion particle. With the coexistence of these super partners, the infinities associated with the issue of Higgs boson interacting with itself and giving mass to itself are taken care of. These massive sparticles could also account for the dark matter.

In Tev-scale energies, supersymmtery calculations do unify the electroweak force with the strong force within 1%, and even more accurate convergence could be expected. Therefore, the unification of the three forces would become a fact if SUSY was proved. SUSY could actually extend beyond unification of the three forces. In specific instances, SUSY has already been applied to quantum gravitational calculations. SUSY is consistent with the so-called String theory – theory of everything, and would not need to be abandoned if the String theory proves to be right. The LHC with its high-energy collisions would look for the super partner particles. Any discovery of a super-particle would be momentous in the history of physics, comparable only to the discovery of electron.

[However, a new experiment conducted to an astounding accuracy of 10^{-27} cm, in May 2011, in Imperial College, London (J.J. Hudson et al, Nature, 473, 493–496 (26 May 2011)), found that electron interaction with an electric field is spherically symmetric. This is possible evidence that SUSY, if correct, will have very restrictive results and, therefore, LHC experiments will have a tougher time proving SUSY.]

Discovery of Other Dark Matter Candidates

LHC will watch for the evidence of neutral and stable matter that could be candidates for dark matter. In addition to the sparticles, the candidates are weakly interacting massive particles (WIMP, just a term thrown in instead of just calling them "a type of dark matter"), axions hypothesized to account for CP violation, and hidden sector particles which interact only through gravity. Such particles would be indirectly observed from the missing energy/mass after a collision event.

Matter–Antimatter Imbalance

The fact that we and the objects we use exist, shows that there is more matter than antimatter. If the Universe had equal amounts of matter and antimatter, as would be expected from most symmetries, then the stable objects such as stars, elements, and molecules would not have originated, and the Universe would be completely busy with pervasive annihilation and pair production processes. Instead, the annihilation is expected to be limited to domains that exclude regions where we observe matter. Annihilations produce a diffuse gamma ray background. Since physicists know the annihilation rates and can measure the distance of the regions that have this background radiation, one can determine the volume of these domains and the matter–antimatter imbalance. Currently, the estimate of the sum of the size of these domains is very close to the size of the Universe, showing that the matter–antimatter imbalance must indeed be small. (Recent Fermilab experiments show that this imbalance might be much higher than expected – as much as 1%, while previous estimates indicated 0.001%.)

In 1967, Andrei Sakharaov proposed that matter–antimatter imbalance can arise out of (a) non-conservation of baryon (neutrons and protons) numbers in order to give rise to baryon–antibaryon imbalance, (b) high rate of CP (combined Charge Conjugation and parity) conservation violation (see Chap. 11), and (c) nonequilibrium thermodynamics. Of these, D. Toussaint, S. B. Treiman, and Frank Wilczek [Phys Rev D, p. 1036 (1979)] show that nonequilibrium thermodynamics during Universal expansion would cause the baryon conservation to be violated, and therefore (c) would be needed to cause (a). The so-called Hawking Radiation from black holes (see below) would be involved in such a process. Currently, there is no evidence for nonconservation of baryons.

Matter–antimatter imbalance can be examined by more accurate observation of CP violation. Currently, the Standard Model does not have the mechanism to bind the observed range of CP violation which is consistent with the amount of antimatter in the Universe. The Quantum Chromodynamic theory in the SM would give no CP violation. This is inconsistent, because this lack of CP violation is accompanied by a prediction of neutron dipole moment which is trillion times larger than the observed value. If the SM is modified with additional postulates, such as axion particles, two time dimensions, etc., this would give massive CP violation. The weak interaction component of the SM, consistent with the observed rate of CP violation, can currently only account for CP violation in an amount of matter in just one galaxy. Given the state of the theory, it is clear that we need more accurate experimental data to pin down the amount of CP violation in nature.

The LHC would be a powerful "B" factory producing a large number of B-mesons (containing b – bottom or beauty – quark) which can decay through a large number of weak interaction paths that can exhibit the CP violation (anisotropic emission of particles in a decay with charge conjugation). Therefore, the matter–antimatter imbalance can be estimated from LHC experiments that lead to a much clearer estimate of the CP violation.

In this experiment, one would measure the rate of a neutral B meson decay into A+ and A− particles (where A are decay particles and the sign refers to their electric charge) and compare this with the rate of decay of anti-B (B-meson and B-bar are produced at the same rate). If Bbar decays faster, then it can account for the antimatter imbalance. Since B-mesons live only for a trillionth of a second, one would need fast-moving B-mesons so that they can be first identified at their birth and then decay can be observed within the detector with sufficient length of tracks.

Quark-Gluon Plasma

Like the Relativistic Hadron Collider in Brookhaven National Lab, the LHC will collide extremely energetic lead ions, a bundle of a large number of quarks. This massive collision will create a soup where, for a substantial moment, no particle identities will exist. The lead ion collisions with center of mass energy of over 2,000 TeV with high luminosity will nearly result in conditions that existed microseconds after the Big Bang, and the collision productions will be similar to the "quark-gluon" state that must have existed soon after the Big Bang. The study of this plasma will reveal the nature of particle creation and indeed how the Universe might have evolved. In this way, this is a time machine to observe the whole Universe and its evolution (Such matter has densities higher than that of neutron stars, where a teaspoon of matter would weigh several tons.).

In May 2011, LHC was already producing such plasmas with temperatures and densities twice that had been created before.

Gravity and Extra Dimensions

Gravity is known to be the weakest force, but the Standard Model has no handle on why that is so. One possibility is (for example, in string theory) that there are more dimensions that are curled up in small distances so that we do not experience these dimensions (like an ant walking across a tightly wound paper cylinder). It is possible that gravity is actually a stronger force that what we experience is what leaks out into our dimension. We have to probe gravity at the level of these dimensions. LHC with the beam of high momentum may be able to do that.

Other Exciting Physics

There are a number of new ideas that are awaiting confirmation or rejection. One example is the theory that quarks are not fundamental particles, but are made of preons. Since fundamental particles do not have excited states, discovery of an excited state of quark (q*) would prove that a quark is not a fundamental particle. The ATLAS detector measured background signals in the 7-TeV collisions in LHC in September 2010 and came up with highly accurate limits on the energy of the q*, by observing the "dijet" events. This has been accepted for publication in the Physical Review, marking a milestone for the flow of experimental results. The

CMS detector has made another exciting discovery by finding a, hitherto unantici-pated, phenomenon of correlation between the angle of emission of particles coming out of a collision. This might point to new symmetries. Many such "off-the-chart" results are anticipated in the LHC.

With its powerful beam of colliding particles and elaborate detectors, the LHC is the first ever machine to examine simultaneously the range of the most fundamental physics questions described above. In the coming years, LHC will do for high-energy physics what the Hubble telescope did for astrophysics and astronomy.

Large Hadron Collider in a Nut Shell

Accelerator Ring

The experimental goal of the LHC is to generate one hadronic event each time the beams cross. The limit is set from the point of view of being able to manage the total number of interesting and uninteresting events, and the data acquisition and analy-sis rates. With the given accelerator parameters and an energy of 7 TeV, this translates to about 2,800 bunches , each with 10^{11} protons, a beam size at collision of 16-µm diameter, and a bunch length of about 7.5 cm. The beams would cross at an angle of about 300 µrads every 25 nanoseconds and collide.

The chain of LHC accelerators (Fig. 13.2) follows previous prescriptions. The protons are born in an ion source and accelerated to 50 MeV in a Linac. They are then accelerated to 14 GeV in the proton synchrotron (PS) booster ring and to 420 GeV in the SPS. The beams would then be accelerated to the full energy of 7 TeV in the LHC ring. The LHC ring consists of 22 km of curved section and eight straight sections, totaling 5 km in length, for injection, beam cleaning, target experiment, beam dumps, and the big detector experiments. Of the curved sections, 80% are filled with 53.5-m-long half-FODO cells. As stated before, the two proton rings are enclosed by a single magnet assembly with two apertures (Figs. 13.3 and 13.4). The ring has 1,232 dipoles magnets and 392 quadrupoles with an aperture of 70 mm. Each arc section is called a sector and magnets in each section are connected in series (~150 magnets). In addition to the protons, the injectors can also supply lead ions and these can be accelerated to 1,300 TeV.

The two parallel beams are brought into a ˙crossing angle at the collision point, using a number of bending dipoles with varying distance between apertures, and focused to get the fine beam of less than 20 micron diameter. Four triplets of quadrupoles are used for the final focusing, each with the strength of 200 T/m and varying in length from 5.3 to 6.5 m.

To enable the LHC magnets to achieve high magnetic fields, all the magnets are cooled with pressurized superfluid liquid helium (at 2 K), which penetrates and fills all gaps, has 100,000 times the heat capacity of the metals, and therefore provides powerful cooling in addition to providing higher superconducting critical current

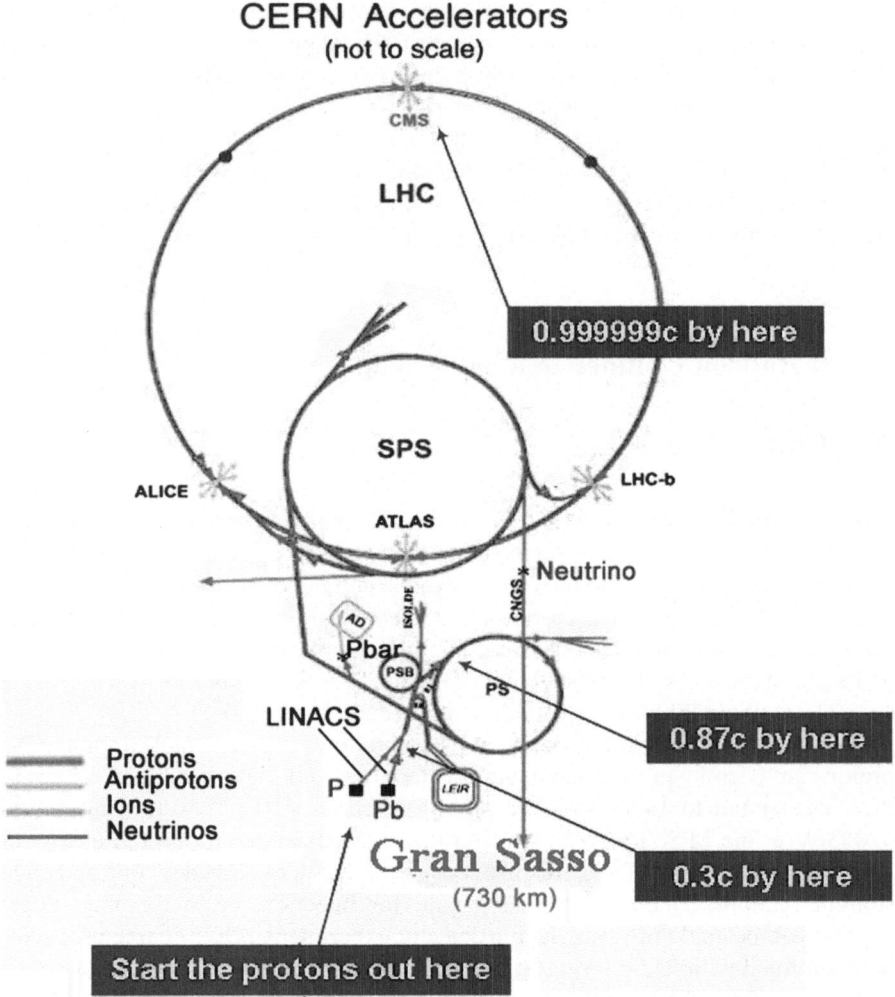

Fig. 13.2 The layout of accelerator stages, experimental detectors, and utility sections in LHC (Courtesy: CERN, European Organization for Nuclear Research)

capacity. The beam tube is maintained at about 10^{-10} mmHg to enable storing of the beam for up to 10 h without loss by scattering with gas molecules. The cryostat vessels are also maintained in high vacuum. The LHC beam is accelerated by eight superconducting cavities in each ring, each producing 5 MV/m at a power of 2 MW and a frequency of 400 MHz. The LHC (LEP) tunnel had been built underground at an average depth of 100 m, and 1.4° gradient was used to reduce the cost of vertical shafts. The choice of the depth was made because it had to be below the green sandstone level and had to match the PS extraction point.

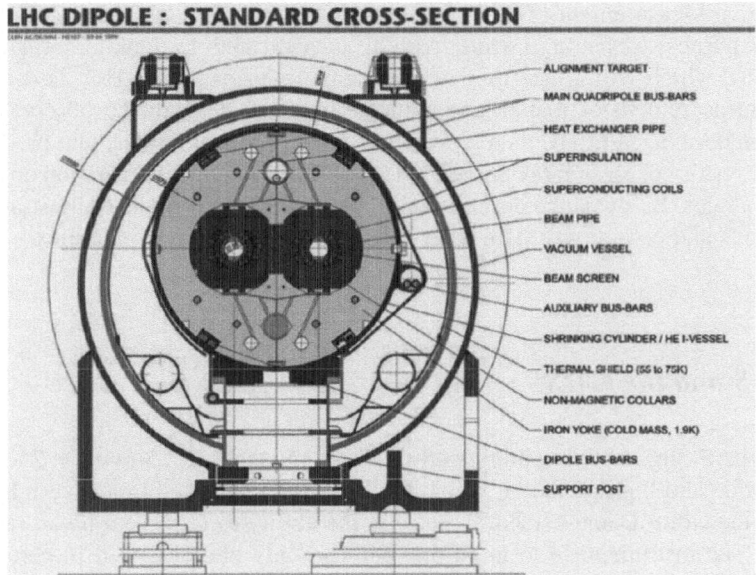

LHC DIPOLE : STANDARD CROSS-SECTION

ALIGNMENT TARGET

MAIN QUADRIPOLE BUS-BARS

HEAT EXCHANGER PIPE

SUPERINSULATION

SUPERCONDUCTING COILS

BEAM PIPE

VACUUM VESSEL

BEAM SCREEN

AUXILIARY BUS-BARS

SHRINKING CYLINDER / HE I-VESSEL

THERMAL SHIELD (55 to 75K)

NON-MAGNETIC COLLARS

IRON YOKE (COLD MASS, 1.9K)

DIPOLE BUS-BARS

SUPPORT POST

Fig. 13.3 The double beam tube (2 in 1) superconducting dipole bending magnet (Courtesy, CERN, European Organization for Nuclear Research, Geneva, Switzerland)

Fig. 13.4 The magnets in the tunnel of LHC (Courtesy, CERN, European Organization for Nuclear Research, Geneva, Switzerland)

The Detectors and Experiments

Continuing the metaphor of the snake eating its tail, the LHC collisions will create the fundamental particles, the finest of matter, and at the same time produce collision products like quark-gluon plasma of the early Universe. To exploit this unified opportunity, LHC has four detectors: ATLAS, the Compact Muon Solenoid (CMS), A Large Ion Collider Experiment (ALICE), and the LHC beauty (LHC[b]), all installed in underground caverns. The ATLAS and CMS will conduct all possible experiments, but their focus will be to look for the Higgs boson and supersymmetry particles. There are two other detectors – a relatively tiny small Detector, the LHC forward (LHC[f]), installed near the CMS, and Total Elastic and

Diffractive Measurement (TOTEM), near the ATLAS. The CMS and ATLAS are general purpose detectors, while ALICE is dedicated to looking at lead ion collisions, which will create the so-called gluon plasma. The LHC[b] is dedicated to the measurement of matter–antimatter asymmetry through the observation of B meson decays, and LHC[f] is a specialized detector to test models that predict the primary energy of cosmic ray particles. TOTEM will provide information on proton collision rates. Below is the description of three of these detectors to illustrate their magnitude and complexity. The other detectors are also equally complex.

ATLAS and the CMS

The ATLAS (Fig. 13.5) is a mighty 48-m (about 150 ft) long, 25-m wide, 25-m high (80 ft wide and high) detector, weighing 7,000 ton, is housed in a cavern half the size of the Notre Dame cathedral, and has the ability to observe a broad range of events. The instrument is so large that one can only photograph it in parts, even given the cavern size. All the specific detector assemblies with 100 million channels are immersed in a magnetic field from a central solenoid which produces 2 T, using eight (horizontal) stacks of 2.4-m diameter superconducting coils which store 1 GJ of energy. The detector consists of an inner tracker to measure the momentum of charged particles, two calorimeters to measure particle energies, and a muon detector assembly. One calorimeter is 6.4 m long, 0.53-m-thick barrel containing

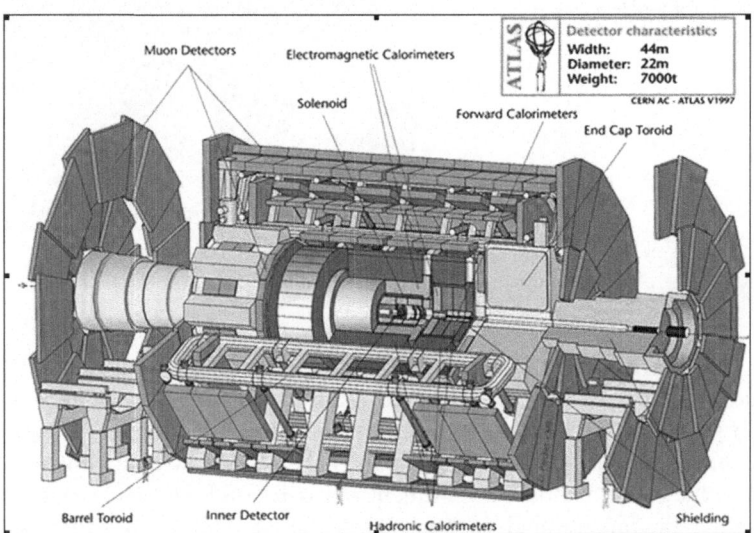

Fig. 13.5 The ATLAS detector (see the size of people in the corners and middle) (Courtesy, ATLAS Detector Collaboration, CERN, European Organization for Nuclear Research, Geneva, Switzerland)

liquid argon at −183°C, and the other is a tile calorimeter with plastic scintillator tiles formed into 64, 5.6-m-long wedge modules. The muon detector has two types of wire chambers (one for fast and another relatively slow but noise tolerant). These wire chambers are designed to trigger on specific range of muon momenta, identifying specific interactions. The muon system also has drift tubes to measure the curvature of tracks of particles bending in the magnetic field and other chambers to measure particle coordinates with an accuracy of 60 μm at the end caps. The inner tracker consists of a pixel detector with 80 million silicon pixels, each 50 × 400 μm. It also has a silicon microstrip tracker with 60 square meters of silicon surface to give a spatial resolution of better than 20 μm in the position of the particle at any instant. There are additional gold-plated tungsten wires – the so-called straw tube tracker- with 400,000 channels.

The CMS detector (Fig. 12.10) has a similar approach to detection with a central tracker, electromagnetic and hadronic calorimeters, and muon detectors, but here the detectors are built around a powerful 6 m diameter and over 12 m long superconducting solenoid which produces a 4-T field, storing 2.7 GJ of energy. CMS weighs 13,000 ton. In an unusual assembly and handling, the detector was assembled on the surface and then lowered into the cavern.

The LHC Beauty

The LHC beauty, referring to bottom quark b-hadrons, is a specialized detector aimed at measuring CP violation (see above). The B-mesons and b-hadrons are produced in a forward cone and this cone is selected by the detector. Two trackers on either side of a 10-m-long magnet help determine the particle momenta. The asymmetry of interactions between the vertical and forward interactions (see Chap. 10) would be measured to quantify the CP violation. An aft RICH (Cerenkov) counter identifies particles and velocities, and there are general and muon calorimeters (Fig. 13.6).

Sociological Impact of the LHC Project

Not surprisingly and to the delight of many physicists, the LHC has attracted popular attention. As is fitting for a venture which is no less exciting and complex than man landing on the moon, the LHC has stirred up people's imagination. Though the content of the program is much too complex for general public, expectations are high from this colossal experiment. This imagination and expectation have two parts: one is a positive view that the new discoveries would change their lives for the better and the other is the sensational speculation which states that the LHC either will fail or is going to bring doom. Not having the ability to attach weights to different arguments and conclusions on the basis of scientific analysis or

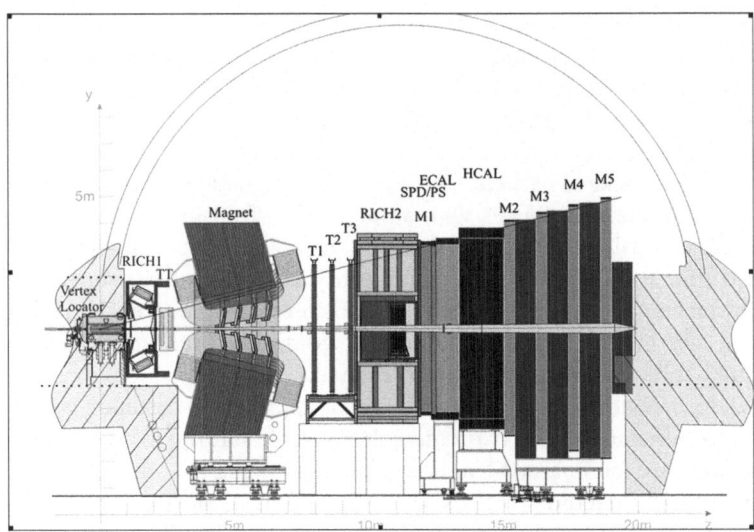

Fig. 13.6 The LHC-beauty detector layout. The p-p collision point is on the left at the Vertex Locator. TT is the main inner tracker (Courtesy, LHCb Collaboration, CERN, European Organization for Nuclear Research, Geneva, Switzerland)

unable to assess the knowledge and credibility of the people who provide such speculative/alternative information, the public takes it all as equally valid. As is natural, if it is sensational, people just eat it up. However, much of the speculation on the potential serious harm that LHC can come to or do is improbable or outright wrong. The safety issues in LHC have been investigated several times by the CERN, and independent calculations and reviews have been made (see for example, CERN web reports: "Safety of the LHC" and "Review of the Safety of LHC Collisions" by LHC Safety Assessment Group with references therein), and these show that none of these speculations rise to the level of potential dangers. Following are some of the speculations:

- LHC will have bad luck
  LHC will self-destruct before finding a Higgs boson, because the discovery is not meant to be: In their papers, Holger Nielsen and Masao Ninomiya (String theorists) argue that nature is abhorrent to the discovery of the Higgs boson and, therefore, the causation will travel back in time to prevent the discovery. Their original paper was "..concerned with looking for backward causation and/ or influence from the future, which, in our previous model, was assumed to have the effect of arranging bad luck for large Higgs producing machines, such as LHC and the never finished SSC (Superconducting Super Collider) stopped by Congress because of such bad luck, so as not to allow them to work," and they speculated that initial conditions might have been arranged for them to fail (See H. Nielson and M. Ninomiya, Int. J. Mod. Phys. A 23: 919–932, 2008, H. Nielson and M. Ninomiya, CERN Document CERN-PH-TH-2008-035; OIQP-07-20;

YITP-07-99, "*A particle God does not want discovered*", Jonathan Leake, The Sunday Times, Oct 18, 2009). There is an interesting hypothesis in this paper, which is not unscientific in its approach. They propose that if the quantity called "action" is taken to be a complex quantity (albeit arbitrarily and contrary to past experience), we get a result that all actions become nonlocal and make certain that trajectories of events will become improbable [In general mechanics, the well-verified law is that action, which is the integral of the Lagrangian (difference between kinetic energy and potential energy) over time, is a minimum for the trajectory that is actually observed. This is the equivalent of stating that the shortest distance between two points is given by a straight line. In simpler terms, Nature is frugal]. By hypothesizing an action that is complex, Nielsen and Ninomiya show that certain small action trajectories, which would have been most probable, end up with low probabilities. They speculate that one such trajectory is the discovery of the Higgs boson. They go on boldly to speculate that nature, in preventing the discovery of the boson, will choose to cause an irreparable accident in LHC. Literally what this means is that the discovery will be so destructive in the future (after discovery) that the Higgs boson will travel back in time and destroy the machine, before the very proton beam collision.

Although this "unorthodox" idea sounds interesting and scientific and is even suggestive of nonlocal quantum effects, the fundamental basis for a complex action is untenable. In the vast experience of physics theory and experiments, there is no evidence for a complex action or for backward causation in time, and neither does it resolve any existing physics issues to make this approach a worthy one. The idea that the Higgs will thwart its own discovery is untestable. Above all, this could have been foretold for the discovery of quark, which did happen and no machine was broken. Given all these, this speculation, even if valid, is extremely improbable and not worth worrying about.

The authors have now revised their approach to say how LHC may be run (using "a card-drawing experiment") so that success may be assured.

- Mini Black Hole

There have been observations that confirm the existence of black holes in the Universe. Black holes are massive star-like objects whose density of matter is so high that their gravity is intense and they pull in any neighboring matter. Space is so curved around them that radiation (light) is bent around them and even light cannot escape. A popular fear was created that LHC with its intense concentration of energy will produce mini black holes which would escape the machine. The scenario would then go on to the mini black hole burrowing through earth, devouring it and then beyond.

Ordinarily, below the Planck energy of 10^{19} eV, one would not expect to produce black holes, but N. Arkani-Hamed, S. Dimopoulos, and G. Dvali (ADD) published a paper in 1998 which showed that if extra dimensions exist as is postulated in the String theory (Russell–Sundrum model), the threshold energy would be reduced to a TeV or so, since gravity increases much faster with decreasing distance in the extra dimensions than the inverse square law observed by matter in 3 dimensions. The particles, we have observed so far, are 3 dimensional, but a high energy of collisions could produce particles that exist in multi-dimensional space, and the ADD model predicts the formation of mini black holes. Other calculations by S. B. Giddings and S. Thomas also show that the black hole formation in LHC collision has a very low but finite probability, and obviously, only if the physics model is valid. But the public commentary on this has centered on the safety of the LHC machine and whether LHC mini black hole will cause a catastrophe.

First, why such a scenario is unlikely: In its 4.5 billion years of existence, the earth has been bombarded by cosmic rays of energy up to 10 million times that of LHC protons. Even granting that the target of the cosmic rays is fixed (meaning it is not a head-on collision of particles), the equivalent energy is up to thousand times that of the LHC beam. In its life, the earth has seen more than 3×10^{22} collisions with such energies, while LHC will create only 10^{17} collisions in its operational lifetime. In the scale of solar system, there have been over 3×10^{26} collisions of high energy, a billion or more times than LHC will observe. Including astronomical bodies such as planets and stars, which also have existed for billions of years, nature has carried on trillion times LHC-type collisions. But the phenomenon of creation of destructive earth-devouring black hole has not occurred in known history of the earth and the solar system, and neither are there any observations of mini black holes in the Universe. All leading scientists like Steven Hawking have assessed the situation and come to the same conclusion. Hawking states, *"The world will not come to an end when the LHC turns on. The LHC is absolutely safe... Collisions releasing greater energy occur millions of times a day in the earth's atmosphere and nothing terrible happens."*

One reason we do not see mini black holes is that all black holes are theorized to evaporate by the so-called Hawking Radiation. This radiation arises because at the horizon of a black hole, beyond which matter and radiation cannot escape, matter–antimatter pair would be produced from gamma rays, as in the vicinity of any other high-energy object. At the horizon, often one of the pair (matter or antimatter) may fall into the black hole, the other of the pair may escape, causing energy loss to the black hole, which is termed as the Hawking Radiation. If born, the LHC mini black holes would have a radius of 10^{-11} cm and temperature of about 10^{16} K, and such a black hole would evaporate within 10^{-24} s, well before it gets out of the collision zone.

But one might argue that Hawking Radiation is not (yet) an observed fact and an LHC black hole which is neutral might not interact, be "slippery," and might travel to the center of the earth. But if such black holes were to be produced and lodged into the earth, the accretion rate of matter would be such that it would take thousands of years. This reasoning, if pursued for dense matter like neutron stars and white dwarfs, demonstrates a fallacy. The long life time of these stars (billion years) precludes the existence of such long-lived mini black holes, because the accretion rate for a black hole in a neutron star would be a few years, and if mini black holes lived long, the neutron stars would have become black holes themselves.

The argument based on cosmic ray experience, much more powerful, and earth and Universe's long exposure also preclude the destructive creation of objects like vacuum bubbles (expanding the lowest energy state of space) and hypothetical magnetic monopoles which might convert ordinary proton into electrons in a chain reaction.

- Strangelets:

We saw in the previous chapter that strange particles with non-integer spin have three quarks of mixed flavors. For example, the lambda baryon has an up, a down, and a strange quark. Such quarks decay into normal matter with short lifetimes. But it is hypothesized that strange particles with more number of quarks and being the size of a nucleus, would be more stable than ordinary nuclei. These "strangelets" are more stable because with the greater variety of quantum numbers, Pauli's exclusion principle (which forbids more than one particle with same quantum numbers to occupy a given state) allows packing of the ground state. If the surface tension of the strangelet nucleus is larger than a critical value, then larger strangelets are even more stable (nuclei have forces similar to liquid drops). Strangelets are candidates for dark matter. However, no strangelets have been detected so far.

If indeed large strangelets exist with greater stability, then ordinary matter coming in contact with strangelets might be converted to strange matter, and a chain reaction might convert the whole earth into hot strange matter. Cosmic ray strangelets, on the way to earth, would decay into positively charged particle and be repelled by nuclei. But a collider could produce negatively charged strangelets which would attract normal nuclei. However, even the Relativistic Heavy Ion Collider, which has a higher probability of producing strangelets than LHC, has not created them.

In summary, the concept of the existence of stranglets has not been given credence by any experimental or astronomical observation. Even if these exist, it is unknown if these are stable. If these exist and are stable, they are more likely to be cold and positively charged (noninteractive). Therefore, all indications point to this being not a safety issue.

As pointed out in the CERN report, the conclusion that the LHC does not pose any serious safety issues is based on sound empirical reasoning and observations and experimental results. Therefore, the conclusions also apply to future speculations on other exotic phenomenon. In a lighter vein, no CERN physicist is planning to relocate to Australia or Canada when LHC achieves full energy.

The Exploration of the Next Frontier Has Begun

Like Robert Wilson of the Fermilab, the scientists and engineers of the world have become "Frontier Men and Women." Though we might be unable to explore all physical frontiers, there would always be frontiers to explore in science. The LHC, with its successful startup, promises to at least give a glimpse of the predicted and unforeseen physics results. Even with the presently achieved collisions, these new revelations have started pouring in. Since the commissioning of LHC, these are some of the new results, that show that we are at a new frontier

- The LEP Collider at CERN results showed that the Higgs boson has to have a rest mass greater than 114 GeV/c^2. The Fermilab Tevatron experiments excluded the range of 158 GeV/c^2 – 175 GeV/c^2. Now, with record collision energies (3.5 TeV on 3.5 TeV), LHC (ATLAS and CMS) results show that the Higgs rest mass cannot be in the range of 150 GeV/c^2 – 205 GeV/c^2. (These are stated with 90–95% confidence level).
- The LHC^b surpassed the precision with which Tevatron investigated matter-antimatter asymmetry using B mesons. LHC^b has discovered that the matter-antimatter imbalance is not as strong as Tevatron results indicated and has set the upper limit of 4 times the prediction of the Standard Model. While this is a positive for the Standard Model, the issue of accurate determination of matter-antimatter symmetry in the Universe will require more work.
- The ATLAS results, though have not found any evidence of supersymmetric particles, but have been able to determine the lower limit on the ratio of the squark mass to gluino (super partner of gluon) mass. This indicates that the super particles, if they exist, would have masses 900 GeV/c^2 or higher. This would mean that they would be produced very rarely and lot of events would have to be recorded before any conclusions can be drawn.
- Using lead ion collisions, the ALICE detector observed that the quark-gluon soup behaves like an ideal fluid.

In 2012, when the machine reaches the full energy, there would be breathless anticipation of results. The coin has been dropped into the vending machine and the child in us is eagerly waiting for the treats. The LHC is the culmination of human endeavor of several centuries in modern science. The shoulders of all the past accelerator and detector builders and theoretical physicists are supporting this endeavor. The effort combines the advances in physics, engineering, mathematical knowledge, the development of many materials from ordinary metals to exotic alloys and polymers, and the latest developments in information technology. Above all, it brings forth the extraordinary spirit of exploration and cooperation that drives us all and the wisdom to direct this spirit in a fruitful journey.

Index

R. Jayakumar, *Particle Accelerators, Colliders, and the Story of High Energy Physics*,   219
DOI 10.1007/978-3-642-22064-7, © Springer-Verlag Berlin Heidelberg 2012